Student Solutions Manual

STATISTICS
A Bayesian Perspective

Donald A. Berry
Duke University

Duxbury Press
An Imprint of Wadsworth Publishing Company
I(T)P® An International Thomson Publishing Company

Belmont • Albany • Bonn • Boston • Cincinnati • Detroit • London • Madrid • Melbourne
Mexico City • New York • Paris • San Francisco • Singapore • Tokyo • Toronto •
Washington

Printed in the United States of America
1 2 3 4 5 6 7 8 9 10

For more information, contact Wadsworth Publishing Company.

Wadsworth Publishing Company
10 Davis Drive
Belmont, California 94002, USA

International Thomson Editores
Campos Eliseos 385, Piso 7
Col. Polanco
11560 México D.F. México

International Thomson Publishing Europe
Berkshire House 168-173
High Holborn
London, WC1V 7AA, England

International Thomson Publishing GmbH
Königswinterer Strasse 418
53227 Bonn, Germany

Thomas Nelson Australia
102 Dodds Street
South Melbourne 3205
Victoria, Australia

International Thomson Publishing Asia
221 Henderson Road
#05-10 Henderson Building
Singapore 0315

Nelson Canada
1120 Birchmount Road
Scarborough, Ontario
Canada M1K 5G4

International Thomson Publishing Japan
Hirakawacho Kyowa Building, 3F
2-2-1 Hirakawacho
Chiyoda-ku, Tokyo 102, Japan

ISBN 0-534-23476-3

To the Student

   This manual contains solutions for the odd-numbered exercises in the text. In this manual I show you how I would solve the exercises. The exercises in the text tend to be very similar to examples in the text. While most of these solutions contain a good bit of detail, the explanations are not usually as detailed as are those in the examples. If you do not understand a particular solution, try referring to an example in the text that is similar to the exercise and relate the various calculations to those in the example.

   Your instructor may ask you to do some of the even-numbered problems for homework. The even-numbered exercises tend to be similar to the odd-numbered exercises, and so following the solution to a similar odd-numbered exercise will usually help you solve an even-numbered exercise. However, there is no perfect correspondence between evens and odds, and so you cannot simply refer to the solution to exercise number 5, say, when you are working on exercise number 6.

   The only odd-numbered solutions not included in this manual are those for exercises that require some individual research on your part. However, in a few of these cases I tell you the results of my research, and then I carry out a solution based on these results; your results and hence your solution will probably be different.

   In some--but not all--of the solutions I use Minitab. In these cases I show the actual Minitab printout. You should get the same answer whether or not you are using Minitab.

   I continue the practice described in the text of carrying as much numerical accuracy as I can and then rounding off to two- or three-digit accuracy in the final answer (except that Minitab output shows greater accuracy). Usually, I show four-digit accuracy during intermediate steps. Suppose you are going through a solution in this manual using your calculator to check the intermediate steps. If your calculator shows the number .12345678, and this is what I got as well, you will see that my version is .1235. Do not change yours to mine, but keep all the digits until the end. Your final answer should be the same as mine.

   A common mistake that students make when doing exercises is using a minus sign. Another is to look up a z-score in the Standard Normal Table and find the area to the left of z instead of the area to the right. Guard against these and similar errors by guessing the answer before you do any calculations. (I made several mistakes of both types in my first draft of these solutions--I should have heeded my own advice!) Suppose you guess 5% probability and then calculate 98%. It is most unlikely that your intuition will be this far off. You probably missed a minus sign or forgot to take one minus the entry in the Standard Normal Table.

Donald A. Berry

TABLE OF CONTENTS

CHAPTER 2
DISPLAYING AND SUMMARIZING DATA

2.1   Dot plot for differences in heights of Darwin's cross-fertilized minus self-fertilized plants (in eighths of inches).

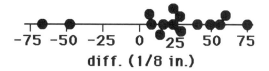

Only two of the 15 differences are negative, indicating that only two of the 15 pots had a taller self-fertilized plant.

2.3   Dot plots of pressures required to remove 16 cannula of type A and 16 of type B:

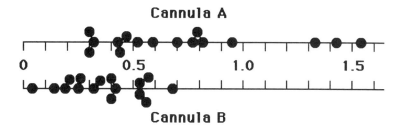

The plots are shown on the same scale for ease of comparison. The dots for Cannula A tend to be to the right of those for Cannula B. So Cannula A usually requires more pressure to remove it.

2.5   Bar chart of improvements in miles walked on drug, rounded from Exercise 2.2:

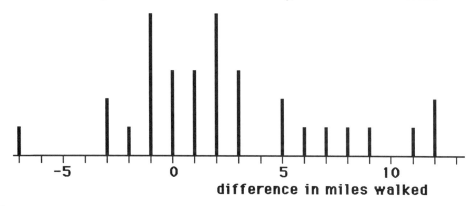

2.7   Bar charts (see next page) for pressures required for 16 Cannula A and 16 Cannula B are shown on same scale for ease of comparison. The vertical scale shown has a slight advantage over a horizontal scale when comparing two histograms. Namely, showing two histograms back-to-back with a horizontal scale requires putting one upside-down in the usual perspective.

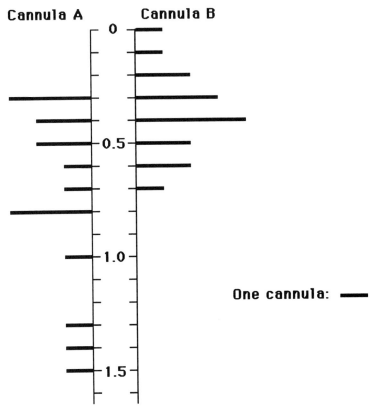

2.9 Scatterplot of pairs of measurements, meas1 and meas2, for 16 hearts:

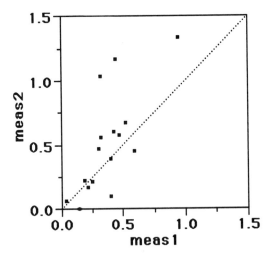

The dashed 45° line indicates observations for which meas1 = meas2 (actually, none of the 16 points lies exactly on the line). The six points below this diagonal indicate that meas1 was greater than meas2 for six of the 16 hearts.

2.11   Scatterplot of diastolic and systolic pressures (mmHg) for pumps A and B:

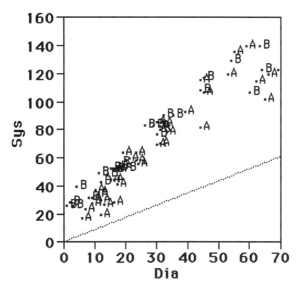

The labelings of A and B sometimes overlap with the dots and with each other. The dashed line shows the points with systolic = diastolic. All points lie above this line because systolic refers to the greatest pressure in the cycle and diastolic refers to the lowest.

2.13

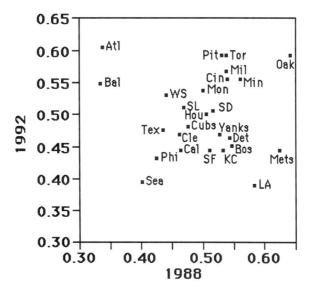

3

The scatterplot on the previous page shows the winning percentage of Major League baseball teams in year 1992 versus in year 1988. As I indicated parenthetically in this exercise, there is little relationship between winning percentages in the two years. Teams that did rather well in 1988 (Oakland, New York Mets and Los Angeles) had records in 1992 that were about the same as teams that did rather poorly in 1988 (Atlanta, Baltimore and Seattle).

2.15 The line plot below shows plasma citrate concentration over time for 10 patients (the dots are labeled with the patient number and the dots are connected).

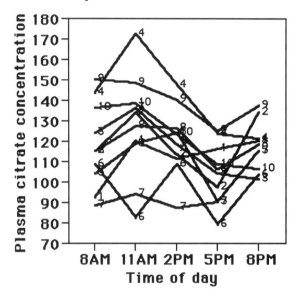

The point that looks most unusual to me is Subject 6 at 11AM. Most other values at 11AM are greater than those at 8AM and also at 2PM. Subject 6 may have had a diet different from the other subjects on the day in question.

[The data disk contains two versions of the data for this exercise: One has the subjects in rows as in the text and the other (called "var") has the times of day in rows. Depending on the way you view the data, one version may be more helpful for making plots than the other.]

2.17 (**a**) Stem-and-leaf diagram of strengths of 50 welds:

```
45 | 0
44 | 5
43 | 0118
42 | 01346
41 | 112469
40 | 0014455688
39 | 0022457888999
38 | 688899
37 | 29
36 | 34
```

4

(b) Histogram of spot-weld strengths using 360-369, 370-379, up to 450-459 as intervals:

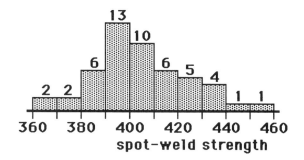

The frequencies are shown atop the bars. These are not required for this exercise but are sometimes helpful. For example, adding the frequencies provides a convenient check that you've not missed something: 2 + 2 + 6 + 13 + 10 + 6 + 5 4 + 1 + 1 = 50, the total number of data points.

2.19 These back-to-back stem-and-leaf diagrams of the diastolic readings allow for easy comparison of the two distributions:

```
            9976 | 0 | 12348
998776665322110 | 1 | 1245678
            4331 | 2 | 069
         9321100 | 3 | 01125
             544 | 4 | 45
             953 | 5 | 4
            9652 | 6 | 034
```

This alternative version with two stems per tens digit is also a correct answer:

```
                 | 0 | 1234
            9976 | 0 | 8
          322110 | 1 | 124
       998776665 | 1 | 5678
            4331 | 2 | 0
                 | 2 | 69
          321100 | 3 | 0112
               9 | 3 | 5
              44 | 4 | 4
               5 | 4 | 5
               3 | 5 | 4
              95 | 5 |
               2 | 6 | 034
             965 | 6 |
```

2.21 (a) The sum of the observations is 146 + 154 + 141 + 140 + 136 + 132 + 147 + 140 + 147 + 139 + 140 + 140 = 1702. Sample mean is then 1702/12 = 141.8.

(b) The sum of the squares is 241,772. So the sample standard deviation is $\sqrt{241772/12 - 141.8^2} = 5.57$. (As a reminder, you'll get a different answer unless you carry at least six significant digits in the sample mean (141.833 and not 141.8).)

5

(**c**) To find the median, order the data:

132, 136, 139, 140, 140, $\boxed{140}$, $\boxed{140}$, 141, 146, 147, 147, 154.

The median is 140, the average of the two middle numbers in boxes.

(**d**) The first quartile is the average of 139 and 140: 139.5. The third quartile is the average of 146 and 147: 146.5.

2.23 (**a**) Histogram of 13 measurements of zinc concentration:

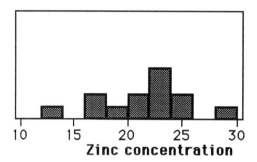

(**b**) To help in finding the median I've repeated the histogram below. This time I've labeled the frequencies and shaded the categories containing the first six observations, that is, those less than 22:

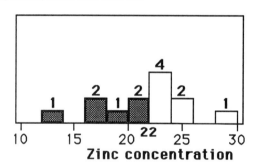

Since n = 13, the median is the next larger observation after the sixth smallest, the smallest one greater than 22, which is 23.3.

(**c**) The sum of the 13 observations is 280.6 and so $\bar{x}$ = 280.6/13 = 21.6. The sum of the squares of the 13 observations is 6269.66 and so s = \r(6269.66/13 - 21.6^2) = 4.05.

2.25 (**a**) Sample mean: The sum of the 32 observations is 77 and so $\bar{x}$ = 77/32 = 2.41, as before (to two-decimal places).

(**b**) Sample standard deviation: The sum of the squares of the 32 observations is 815 and so s = $\sqrt{815/32 - 2.41^2}$ = 4.44.

(c) To help in calculating the median I've repeated the bar chart from Exercise 2.5 below. I've added the cumulative frequencies across the top—these are the numbers of observations to the left of and including the corresponding difference in miles walked. Because n = 32 we want to find the 16th and 17th largest. These are between 15 and 20 and so have a difference in miles walked of +2, which is then the median.

To see this another way, I've filled in the bars for the 16 smallest differences and left them open for the largest 16. The former include one -7, two -3s, one -2, et cetera, on up to one +2. Where the bars switch from filled in to open (+2) is the median.

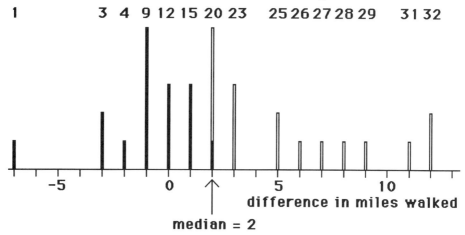

Cumulative frequencies, from left to right:

2.27 (a) Since the sum of the x's is 11.69, $\bar{x}$ = .731.

(b) The sum of the squares is 10.95. So $s = \sqrt{10.95/16 - .731^2}$ = .388. [Remember that I carry more digits than I show.]

(c) Since n = 16, the median is the average of the 8th and 9th biggest, shown in bold face below:

.30	.30	.32	.43	.44	.47	.52	**.59**
.70	.77	.79	.81	.95	1.33	1.43	1.54

So the median is .645.

(d) The first quartile is .43, the median of the first half of the data (omitting the overall median):

| .30 | .30 | .32 | **.43** | .44 | .47 | .52 |

The third quartile is .95, the median of the second half of the data (omitting the overall median):

| .77 | .79 | .81 | **.95** | 1.33 | 1.43 | 1.54 |

2.29

Group:	0	1-2	3-5	6+
Sample mean:	289.4	211.7	318.5	289.4
Standard deviation:	171.9	137.9	226.8	167.7
Median:	210	210	260	260

CHAPTER 3
DESIGNING EXPERIMENTS

3.1 (**a**) There are several ways to do this. Here are two:

Method 1: Decide on route A or B for each day in the first week and then use the other route on that day of the second week. On each day of the first week, toss a coin to decide whether it's going to be route A or B. Once either A or B has been taken three times in the first week, use the other route for the remaining times.

Method 2: Select randomly from the following set of 20 possibilities (by writing them on slips of paper and placing them in a hat, say). These indicate your routes in the first week in order of Mon, Tues, Wed, Thurs, Fri:

AABBB	ABABB	ABBAB	ABBBA	BAABB
BABAB	BABBA	BBAAB	BBABA	BBBAA
BBAAA	BABAA	BAABA	BAAAB	ABBAA
ABABA	ABAAB	AABBA	AABAB	AAABB

This list exhausts the possibilities. I've listed them so that the top and bottom can be thought of as pairs; the nonselected member of each pair is what you'll do in the second week. For example, if you select the third sequence, ABBAB, you'd take route A on Mon and Thurs of the first week and Tues, Wed, and Fri of the second week.

(**b**) There is no one, natural analysis. But because of the possibility of similarities within days of the week it seems reasonable to analyze the five differences in travel time, route A minus route B, for Mon, Tues, Wed, Thurs, and Fri.

3.3 Consider each of the 17 valves in turn. For each of these 17, randomly select one or more valves from the same lot. These will serve as controls. Presumably, these controls did not leak, but you should make sure of that. Make the six measurements on all valves, including the 17 leakers and the controls. Comparing these measurements, leakers vs controls, may help you see what aspect of the production gives rise to leaking. Of course, it may not; this is a complex problem that may not have an easy solution.

3.5 It is nearly impossible to attribute cause from an observational study, regardless how strong the connection. As indicated in Example 3.1, to show that there is a tendency for two characteristics to coexist is not to show that one causes the other. There may be other attributes that give rise to both. One obvious possibility is that people suffering from PTSD (as opposed to requiring other types of medical care) may be more willing to talk about problems stemming from their childhood.

At most these findings indicate that people experiencing PTSD report child abuse more than others needing medical help. Conclusions about cause require a believable explanation and not simply a relationship.

3.7   The spouse of someone who is highly paid has more flexibility than the spouse of someone who is less well paid. While some spouses of both types will choose to have a career, it's likely that more of the first type will not. This study is an instance in which the attributed effect (higher salary) is more likely to be the cause.

3.9   There are clear biases in Helms's sample. Consider the nonresponders, those who did not choose to write to Helms. They likely have very different views than do the responders. Indeed, people are more likely to communicate their views to others with similar views. And people are more inclined to complain about a circumstance than to voice support for it. It is difficult to imagine military people writing to *anyone* to voice support for the President, and all the more difficult to imagine someone writing to Helms to voice such support. (Helms may or may not have realized that the evidence he was presenting was irrelevant.)

3.11 One possibility: Identify villages that are likely candidates for the disease. Logistical and resource limitations restrict the number of villages to say ten. Randomize, assigning five to vaccine and five to a placebo vaccine. Keep both residents and health officials ignorant about the actual assignments. Within each village, every member is assigned the same intervention, either vaccine or placebo.

Analyze the rates of disease within the two groups. The analysis may be complicated, but the analyst should pay particular attention to whether people who develop cholera in vaccine-assigned villages were actually vaccinated. Also important is the prevalence of disease in neighboring villages that were not among the ten in the randomized study.

3.13 One type of multiplicity: Disease characteristics such as estrogen receptors, tumor size, and stage of disease. Another type of multiplicity: Diet characteristics such as total fat, saturated fat, carbohydrates, etc. The fact that the investigators found something of interest in one of the many possible combinations of multiplicities is hardly surprising. This diminishes the report's credibility.

3.15 Parallel: Two possibilities: (i) Select a random sample of women and another of men, and (ii) Select a single random sample from a mixed population, but one in which the men and women are identified.

Paired: There are many possibilities for selecting randomly from man/woman pairs, including husband/wife couples and brother/sister pairs.

3.17 One way: Use 30 separate pots, one plant per pot and assign them to cross- of self-pollination (randomly).

3.19 In a crossover, half the patients (that is, 16 of them) would be given no drug in the first week and drug in the second week, just as for all 32 in the actual study, and the other half would receive drug in the first week and then no drug.

The point of the crossover is that there may be a period effect (after patients workout in the first week they may tire or lose enthusiasm for the study, or they may be in better shape and able to walk further in the second week) or a placebo effect—one associated with either period.

3.21 Randomly choose 4 of the 24 to be assigned to each one of these 6 sequences:

ABC, ACB, BAC, CAB, BCA, CBA

3.23 Assign 2 plots to each combination of fertilizer and corn variety as follows:

Numbers of plots per combination			
	Fertilizer:		Low
Med	High		
Corn variety 1:	2	2	2
Corn variety 2:	2	2	2

3.25 Assign 200 patients each to the four possible treatment combinations, as follows:

Numbers of patients			
	Vitamin E:	Yes	No
Deprenyl	Yes:	200	200
	No:	200	200

CHAPTER 4
PROBABILITY AND UNCERTAINTY

4.1 (**a**) {1,3,5}

(**b**) {5,6}

(**c**) {3,4,5}

(**d**) {5}

(**e**) {1,2,3,4,6}

(**f**) {1,3,4,5}

(**g**) {5,6}

(**h**) {2,4,6}

(**i**) {2,6}

4.3 (**a**) 3/6 = 1/2

(**b**) 5/6

(**c**) 4/6 = 2/3

(**d**) 4/6 = 2/3

(**e**) 3/6 = 1/2

4.5 (**a**) 1/3

(**b**) 1/3

(**c**) 1/3

(**d**) 1/6; there are 6 possible ways of giving out the cod, trout, and tuna (XYX, XZY, YXZ, YZX, ZXY, ZYX)—in only one them (XYZ) is everyone happy.

(**e**) 0; if two people get what they ordered then so does the third.

4.7 For example, the player picks a 3: In each roll, she loses by rolling a 1, 2, 4, 5, or 6—that's 5 ways out of 6 to lose. Thus there are 5x5x5 = 125 ways out of 6x6x6 = 216 to lose. (Think of two tree diagrams, one with 5 branches at each of three levels and the other with 6 branches at each of three levels.) The probability that the player loses her $1 is 125/216.

4.9 4/47

4.11 (**a**) P(odd) = 4/8

(**b**) P(bigger than 5) = P(6,7,8) = 3/8

(**c**) P(a or b) = P(1,3,5,6,7,8) = 6/8

(**d**) P(~b) = 1-P(b) = 5/8

4.13 1/4

4.15 To two decimals, 1511/10000 = .15.

4.17 Answer depends on the results of your coin-tossing experiment.

4.19

Horse Number	"Probability"	Track Odds Against	Track Pays for $2 Bet
7	.4348	1.3	4.60
6	.5882	.7	3.40
8	.0147	67.2	136.40
9	--	--	--
4	.0203	48.3	98.60
5	.0311	31.2	64.20
2	.1000	9	20.00
1	.0167	59	110.00
3	.0162	60.6	123.20
10	.0075	132.1	266.20
	1.2295		

The track pays $(2 plus twice the odds against) for $2 bet, or $4.60 in the case of Easy Goer. The percentage that the track takes is 1 - 1/1.2295 = 18.66%. Therefore, the dollar amount the track takes for each $1,000,000 in the handle is $186,600.

4.21 Possible answer: I asked my mother and this was our conversation:

Q: Mom, which has more people, California or Canada?
A: I don't know, California?
Q: I want to know how likely it is for you that California has a larger population than Canada. Call B the event that California is larger than Canada. What is your probability of B?
A: I don't know that either. I don't even know what "probability" means.
Q: Okay. I'm going to give you a choice between two lotteries. Suppose you get $100 if you win the one you choose. Would you prefer to receive $100 if B is true (that is, if California has more people) or if I get a red M&M when I select one randomly from this bowl? [I show her a bowl with one red M&M and one green M&M.]
A: I'd choose B.
Q: Since you prefer B to an event whose probability is 1/2, your P(B) > 1/2. Now consider this bowl [showing one with 3 reds and 1 green]. Would you rather have the $100 if California is bigger or if I get a red M&M from this bowl?
A: Now I'd rather have the $100 if you get a red M&M.
Q: Okay, that means your P(B) < 3/4; combined with before, we know 1/2 < P(B) < 3/4. Now consider this bowl [showing one with 5 reds and 3 greens].
A: Now I prefer B.

Q: That means your P(B) > 5/8, and so 5/8 < P(B) < 3/4. Now consider this bowl [showing one with 11 reds and 5 greens].

A: I can't decide between B and red. Look, are you serious about this $100 or what? If not then find something better to do, such as your homework!

I concluded that my mother's probability of B is between 5/8 and 3/4.

4.23 Questions with possible answers are as follows:

Q: Doctor, I want to know how likely it is for you that I will survive the next year. Call this event S. I'm going to give you a series of choices. Please try to give answers that reflect your opinions, imagining that you will receive a very substantial prize for choosing correctly if that will facilitate your decisions. For you, is it more likely that I will survive the next year or that I get a red chip when I select one randomly from this bowl containing one red chip and one green chip?

A: S is more likely. [Whew!]

Q: Since S is more likely for you than an event whose probability is 1/2, your P(S) > 1/2. Now consider a new bowl, showing one with 3 reds and 1 green. Is S more likely or less likely than obtaining a red chip when selecting from this bowl?

A: More likely.

Q: That means your P(S) > 3/4. Now consider this bowl, showing one with 7 reds and 1 green.

A: This comparison is really quite difficult for me, but red is now a little more likely than S.

Q: That means your P(S) < 7/8, and so 3/4 < P(B) < 7/8. Now consider this bowl, showing one with 13 reds and 3 greens.

A: This comparison is again hard for me. Now I choose S.

I concluded that the doctor's probability of S is between 13/16 and 14/16 (or 7/8).

CHAPTER 5
CONDITIONAL PROBABILITYAND BAYES' RULE

5.1 You've already bet $55 on Oakland. Suppose now you bet X dollars on San Francisco.
If Oakland wins the Series, your net gain will be 25 - X.
If San Francisco wins, your net gain will be X(16/5) - 55.
To have the same net gain in either case, these two must be equal.
That is, 25 - X = X(16/5) - 55. Or, X(21/5) = 80. Or X = 400/21.
To the nearest cent, X = 19.05 and 25 - X = 5.95.
So if you bet $19.05 on San Francisco, your net winnings are $5.95 regardless of who wins the Series.

5.3 (a) P(a blue chip) = 12/20 = 3/5

(b) P(a chip numbered 2) = 5/20 = 1/4

(c) P(a blue chip or a chip numbered 2) = 14/20 = 7/10

(d) Joint probabilities are products of individual probabilities:
P(B) = 3/5 and P(Y) = 2/5
P(1) = P(2) = P(3) = P(4) = 1/4
P(B and any particular number) = (3/5)(1/4) = 3/20, as in the table
P(Y and any particular number) = (2/5)(1/4) = 2/20, as in the table

5.5 (a) 25/100 = 1/4

(b) 20/90 = 2/9

5.7 The probability that any one fails is .01. The probability that all four fail is $(.01)(.01)(.01)(.01) = (.01)^4$ = .00000001, or 1 E-8.

5.9 Let A = "the first child is a boy" and B = "the second child is a boy"

$P(A \cap B|B) = P(A \cap B)/P(B) = (1/4)/(1/4+1/4) = 1/2$

5.11 (a) Just pick one case in which probabilities don't multiply: $P(G1 \cap B2) = 0$, but $P(G1)P(B2) = (1/2)(1/2) = 1/4 \neq 0$. Thus the sexes of the two children are not independent.

(b) P(G1), P(G2), P(B1) and P(B2) all equal 1/2.

(c) P(G1|G2) = 1

(d) P(D|C) = 1

5.13 Using multiplication rule: To survive he must survive the first "play", which has probability 5/6, and then survive the second, which has probability 4/5 given that he survives the first. So P(survive) = (5/6)(4/5) = 4/6 = 2/3. And P(loses) = 1 - 2/3 = 1/3.

An easier way: In effect he is going to sample 2 of the 6 cylinders. Each has probability 1/6. So the probability of losing is 2/6.

5.15 P(different signs|random) = $\dfrac{8}{12}\dfrac{7}{12}\dfrac{6}{12}$ = $\dfrac{7}{36}$ = .194

5.17 **(a)** Law of total probability: P(rain tomorrow) =
P(rain both days) + P(rain tomorrow but not today) = 1/5 + 1/10 = 3/10

(b) Definition of conditional probability:

$$\text{P(rain tomorrow|rain today)} = \frac{\text{P(rain both days)}}{\text{P(rain today)}} = \frac{1/5}{3/10} = \frac{2}{3}$$

5.19 Let G1 stand for first chip is green and G2 stand for second chip is green. Law of total probability:

P(G2) = P(G2|G1)P(G1) + P(G2|~G1)P(~G1) = (2/5)(2/5) + (2/5)(3/5) = 2/5

The result is the same as in the previous exercise. The two chips selected are again exchangeable, and now they're independent as well.

5.21 Let H stand for heads and T stand for tails. According to the law of total probability:

P(win) = P(win|H)P(H) + P(win|T)P(T) = (1/3)(1/2) + (3/6)(1/2) = 5/12

Alternatively, the winning branches are shown in bold face on the following tree; the total probability of winning is 5/12.

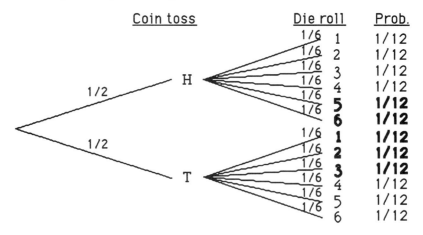

5.23 **(a)** $P(B|A) = P(A \cap B)/P(A) = \dfrac{8/20}{12/20} = \dfrac{2}{3}$

(b) $P(B|{\sim}A) = P({\sim}A \cap B)/P({\sim}A) = \dfrac{1/20}{8/20} = \dfrac{1}{8}$

(c) $P(B) = P(B|A)P(A) + P(B|{\sim}A)P({\sim}A) = \dfrac{2}{3}\dfrac{12}{20} + \dfrac{1}{8}\dfrac{8}{20} = \dfrac{9}{20}$

5.25 If B rolls a 6, B wins with probability 1.

If B rolls a 5, B wins only if both of A's rolls are 5 or less:
Five possibilities for each of two dice: $(5/6)^2$

If B rolls a 4, B wins only if both of A's rolls are 4 or less:
Four possibilities for each of two dice: $(4/6)^2$

If B rolls a 3, B wins only if both of A's rolls are 3 or less:
Three possibilities for each of two dice: $(3/6)^2$

If B rolls a 2, B wins only if both of A's rolls are 2 or less:
Two possibilities for each of two dice: $(2/6)^2$

If B rolls a 1, B wins only if both of A's rolls are 1:
One possibility for each of two dice: $(1/6)^2$

B's score	Prob. B's score	Prob. B wins	Product
6	1/6	1	1/6
5	1/6	$(5/6)^2$	$(1/6)(5/6)^2$
4	1/6	$(4/6)^2$	$(1/6)(4/6)^2$
3	1/6	$(3/6)^2$	$(1/6)(3/6)^2$
2	1/6	$(2/6)^2$	$(1/6)(2/6)^2$
1	1/6	$(1/6)^2$	$(1/6)(1/6)^2$
Sum	1	irrelevant	91/216

Total probability that B wins is 91/216.

5.27 Start over with the probabilities updated as in the previous exercise: $P(A|G) = 0$, $P(B|G) = 1/4$, $P(C|G) = 3/4$. Now use GG to indicate that these probabilities are calculated based on a second G as well as the first G:

$$P(A|GG) = \dfrac{(0)(1/6)}{(0)(1/6)+(2/5)(1/4)+(4/5)(3/4)} = 0,$$

$$P(B|GG) = \dfrac{1/10}{7/10} = \dfrac{1}{7}, \quad P(C|GG) = \dfrac{6/10}{7/10} = \dfrac{6}{7}$$

5.29 Bayes' rule: $P(A|B) = \dfrac{P(B|A)P(A)}{P(B)} = \dfrac{(8/12)(12/20)}{9/20} = \dfrac{8}{9}$

Direct check: $P(A|B) = P(A \cap B)/P(B) = \dfrac{8/20}{9/20} = \dfrac{8}{9}$

5.31 **(a)** $P(S) = 10/50 = 1/5$

(b) $P(S) = (4/5)(5/50) + (6/20)(20/50) + (0/25)(25/50) = 10/50 = 1/5$

(c) $P(E|S) = 4/10 = 2/5$

(d) $P(E|S) = P(S|E)P(E)/P(S) = (4/5) \times (5/50)/(10/50) = 4/10 = 2/5$

5.33 $P(C|+) = \dfrac{P(+|C)P(C)}{P(+|C)P(C)+P(+|\sim C)P(\sim C)}$

$= \dfrac{.99\ \dfrac{1}{1000000}}{.99\ \dfrac{1}{1000000} + .01\ (1 - \dfrac{1}{1000000})} = .000099$

5.35 Data = guard says "Carol" (but wouldn't name you in any case)

Consider the three possible pairs of people to be hanged: YC, YL, LC. Your prior probabilities are 1/3 on each. The event "you get hanged" is the union YC $\cup$ YL. This has probability 2/3 without conditioning on Data.

Assume that if it's LC then the guard would say "Carol" with probability 1/2.

$P(YC|Data) = \dfrac{P(Data|YC)P(YC)}{P(Data|YC)P(YC)+P(Data|YL)P(YL)+P(Data|LC)P(LC)}$

$= \dfrac{1\ (1/3)}{1\ (1/3) + 0\ (1/3) + (1/2)\ (1/3)} = 2/3$

$P(YL|Data) = \dfrac{0\ (1/3)}{1\ (1/3) + 0\ (1/3) + (1/2)\ (1/3)} = 0$

$P(CL|Data) = \dfrac{(1/2)\ (1/3)}{1\ (1/3) + 0\ (1/3) + (1/2)\ (1/3)} = 1/3$

So the probability of You is still 2/3; the information the guard gave is independent of YC $\cup$ YL. But it is not independent of YL $\cup$ CL, which was 2/3 without conditioning on Data and is now 1/3. Changing places with Luke would halve your posterior probability of being hanged to 1/3.

The assumption that the guard says "Carol" with probability 1/2 when Luke and Carol are to be hanged is critical. If this were changed to 1 then the probability of YC drops to 1/2, which is your posterior probability of being hanged. On the other hand, if it is changed to 0 (guard always says Luke when it is going to be Carol and Luke) then the probability of your being hanged *increases* to 1.

5.37 Law of total probability: P(Green on next|Green on first two) =

$$\frac{1}{5}\frac{1}{55} + \frac{2}{5}\frac{4}{55} + \frac{3}{5}\frac{9}{55} + \frac{4}{5}\frac{16}{55} + \frac{5}{5}\frac{25}{55} = \frac{225}{275} = \frac{9}{11}$$

5.39 B = "Man's blood matches that on the glass"
G = "man is guilty"

P(B|G)=1, P(B|~G)=1/100,000

(a) Bayes factor in favor of guilt = P(B|G)/P(B|~G) = 100,000

(b)

P(G):	.00001	.0001	.001	.2	.5	.99
P(G|B):	.5000	.9091	.9901	.99996	.99999	.9999999

5.41 P(I) = .3, P(~I) = .7, P(FF|I) = .5, P(FF|~I) = .25

$$P(I|FF) = \frac{P(FF|I)P(I)}{P(FF|I)P(I)+P(FF|~I)P(~I)} = \frac{.5(.3)}{.5(.3)+.25(.7)} = .46$$

CHAPTER 6
MODELS FOR PROPORTIONS

6.1 Depends on the telephone book you use.

6.3
Prop. W's:	.30	.35	.40	.45	.50	.55	.60	.65	.70
Probability:	.00	.01	.02	.03	.09	.40	.25	.15	.05
Product:		.0000	.0035	.0080	.0135	.0450	.2200	.1500	.0975	.0350

P(we win the next deal)
= .0000 + .0035 + .0080 + .0135 + .0450 + .2200 + .1500 + .0975 + .0350 = .5725

6.5 Depends on the setting and prior probabilities that you chose.

6.7 Using a computer with Minitab, this is the output of p_disc, with the answers to
(a), (b) and (d) of this exercise shown in bold face:

```
DATA> 7 3

    ROW     p    prior   P_x_PRIO   LIKE      PRODUCT      POST     P_x_POST

     1     .1     .01      .001    .000009   .0000001    .000000    .000000
     2     .2     .01      .002    .000786   .0000079    .000044    .000009
     3     .3     .01      .003    .009002   .0000900    .000503    .000151
     4     .4     .01      .004    .042467   .0004247    .002373    .000949
     5     .5     .31      .155    .117188   .0363281    .203017    .101508
     6     .6     .15      .090    .214991   .0322486    .180219    .108131
     7     .7     .25      .175    .266828   .0667070    .372787    .260951
     8     .8     .20      .160    .201327   .0402653    .225019    .180016
     9     .9     .05      .045    .057396   .0028698    .016038    .014434
    10                     .635                                     .666149
```

The answer to (c) is the sum of the four posterior probabilities for p > .5.

Summarizing:

(**a**) .203

(**b**) .016

(**c**) .180 + .373 + .225 + .016 = .794

(**d**) .666

6.9 $s = 2$, $f = 10$, Likelihood $= p^s(1-p)^f = p^8(1-p)^2$

(a)
Model (p)	1/2	5/8	6/8	7/8
Prior prob.	.7	.1	.1	.1
Likelihood	.000977	.003274	.006257	.005369
Posterior prob.	.315	.151	.288	.247

(The probability of the null hypothesis of "just guessing," $p = 1/2$, has dropped somewhat because the evidence supports the hypotheses of some ability more than it does that of the null. But 31% probability of "just guessing" is still substantial, indicating that the evidence is not conclusive.)

(b) P(predict correctly) $= .315 \times (1/2) + .151 \times (5/8) + .288 \times (6/8) + .247 \times (7/8) = .683$

6.11 Using a computer with Minitab, this is the output of p_disc:

```
DATA> 2 23

ROW    p    prior   P_x_PRIO   LIKE      PRODUCT      POST      P_x_POST

 1    .0    .08     .000      .000000   .0000000   .000000    .000000
 2    .1    .12     .012      .265888   .0319066   .841485    .084148
 3    .2    .08     .016      .070835   .0056668   .149454    .029891
 4    .3    .04     .012      .007390   .0002956   .007796    .002339
 5    .4    .12     .048      .000379   .0000455   .001200    .000480
 6    .5    .28     .140      .000009   .0000025   .000066    .000033
 7    .6    .12     .072      .000000   .0000000   .000000    .000000
 8    .7    .04     .028      .000000   .0000000   .000000    .000000
 9    .8    .04     .032      .000000   .0000000   .000000    .000000
10    .9    .04     .036      .000000   .0000000   .000000    .000000
11   1.0    .04     .040      .000000   .0000000   .000000    .000000
12                  .436                                      .116891
```

(a) His posterior probabilities are shown in bold-faced type in the above table.

(b) His predictive probability that the next lined nest is raided is .117, which is also shown in bold-faced type in the table.

6.13 The table on the next page shows the likelihoods. Since only ratios of likelihoods matter and not the likelihoods themselves, I have used a tack that is slightly from the one in the text. Namely, I divided each likelihood by the largest likelihood in the table. This convention serves to get rid of powers of 10 that can complicate tables. It happens that the biggest likelihood occurs at $p = .2$ and is $(.2)^{28}(.8)^{99} = 6.835E{-}30$. Dividing by this in the likelihood column means that the revised likelihood for $p = .2$ is 1 and all other likelihoods are smaller than 1.

(a) Since the prior probabilities are equal, just as for the calculations for Dr. X in Section 6.5, the posterior probabilities are simply the likelihoods divided by their total (1.155) and are shown in bold-faced type in the table.

(b) The predictive probability of a serious reaction within 18 months for the next person in this population who receives angioplasty is shown in bold-faced type in the table: about 21%.

p	Prior	Likelihood	Posterior	p x Post
.0	1/11	.0000	**.0000**	.0000
.1	1/11	.0004	**.0004**	.0000
.2	1/11	1.0000	**.8657**	.1731
.3	1/11	.1547	**.1339**	.0402
.4	1/11	.0001	**.0001**	.0000
.5	1/11	.0000	**.0000**	.0000
.6	1/11	.0000	**.0000**	.0000
.7	1/11	.0000	**.0000**	.0000
.8	1/11	.0000	**.0000**	.0000
.9	1/11	.0000	**.0000**	.0000
1.0	1/11	.0000	**.0000**	.0000
sum	1	1.1552		**.2134**

6.15 $s = 2$, $f = 10$. Likelihood $= p^s(1-p)^f = p^2(1-p)^{10}$

Prop B's:	.1	.2	.3	.4	.5
Prior prob:	.1	.1	.1	.1	.6
Likelihood:	.00349	.00429	.00254	.00097	.00024
Post prob:	.273	.337	.199	.076	**.115**

The posterior probability of the null hypothesis is .115, shown in bold-faced type in the table. (The probability that there is no difference between these two types of bases in preventing injuries has dropped from 60% to about 11% while the probability that breakaway bases are better has increased from 40% to about 89%. An interesting aspect of the table is that if they are better then they seem to be a lot better, with most of the posterior probability being associated with small values of p.)

CHAPTER 7
DENSITIES FOR PROPORTIONS

7.1 Output from Minitab (exec 'p_disc'), with answers in bold-faced type:

```
DATA> 6 0

ROW    p    prior   P_x_PRIO   LIKE     PRODUCT    POST     P_x_POST

1    .500   .500   .250000    .01563   .007813   .035737   .017869
2    .625   .125   .078125    .05960   .007451   .034082   .021301
3    .750   .125   .093750    .17798   .022247   .101767   .076325
4    .875   .125   .109375    .44880   .056099   .256619   .224542
5   1.000   .125   .125000   1.00000   .125000   .571795   .571795
6                  .656250                                 .911832
```

(a) .0357   **(b)** .9118

7.3 **(a)** The posterior density is beta(4, 25).

    (b) The predictive probability is 4/29 = .138.

7.5 $r = .4$, $r^+ = .45$, $s = 7$, $f = 3$

The predictive probability of "in favor" is $\dfrac{a+s}{a+s+b+f}$, but you do not know a and b. To find them:

$$a = \frac{r(1-r^+)}{r^+ - r} = 4.8, \quad b = \frac{(1-r)(1-r^+)}{r^+ - r} = 6.6$$

So, predictive probability $= \dfrac{a+s}{a+s+b+f} = \dfrac{4.8+7}{4.8+7+6.6+3} = \dfrac{11.8}{21.4} = .55$

7.7 Depends on the results of rolling the thumb tack.

7.9 For the beta(19, 4) density, $r = a/(a+b) = 19/23$, $r^+ = 20/24$, and $t = \sqrt{r(r(r^+ - r))} = \sqrt{.005986} = .0774$. The z-score corresponding to proportion .5 is

$$z = \frac{.5 - 19/23}{.0774} = -4.21$$

According to the Standard Normal Table, the probability to the right of -4.21 is 1.0000. (The exact probability to the right of .5 for the beta(19, 4) density calculated using Minitab is .9996.)

7.11 The posterior density is beta(4, 25).

$$z = \frac{.5 - 4/29}{\sqrt{(4/29)(5/30 - 4/29)}} = \frac{.3621}{.0630} = 5.75$$

From the Standard Normal Table the probability to the left of 5.75 is 1.0000. (The exact probability calculated using Minitab is .999986.)

7.13 Posterior density is beta(1+11, 1+8) = beta(12, 9)

$$z = \frac{.5 - 12/21}{\sqrt{(12/21)(13/22 - 12/21)}} = \frac{-.0714}{.1055} = -.68$$

The probability to the right of z = -.68 is about .75

7.15 **(a)** The posterior density is beta(a+s, b+f) where

a = 2, b = 18, s = 44, f = 223

$$r = \frac{a+s}{a+s+b+f} = \frac{46}{287} = .160, \quad r^+ = \frac{a+s+1}{a+s+b+f+1} = \frac{47}{288} = .163,$$

$$t = \sqrt{r(r^+ - r)} = .0216, \quad z_{95} = 1.96$$

The 95% posterior probability interval is .160 ± 1.96x.0216, or from about 11.8% to 20.3%.

(b) A difference between left handers and right handers is supported since the null proportion 10% is not in the 95% posterior probability interval in part (a).

(c) To find the posterior probability that the population proportion is greater than 10%, calculate

$$z = \frac{.10 - .160}{.0216} = -2.79$$

According to the Standard Normal Table, the probability to the right of z = -2.79 is about .997.

7.17 a = b = 1, s = 106, f = 557

$$r = .1609, \quad r^+ = .1622, \quad t = .0142, \quad z_{95} = 1.96$$

The 95% posterior probability interval for the population proportion who would have answered "yes" is .161 ± 1.96x.0142, or from about 13.3% to 18.9%.

7.19 The posterior density is beta(a+s, b+f) where a = b = 1, s = 322, f = 428. For this density,

$$r = \frac{a+s}{a+s+b+f} = .4295, \quad r^+ = \frac{a+s+1}{a+s+b+f+1} = .4303, \quad t = \sqrt{r(r^+ - r)} = .0180$$

(a) $z_{95} = 1.96$; the 95% posterior probability interval is from about 39.4% to 46.5%.

(b) With 4 times the sample size the width of the interval in your answer in (a) would be divided by $\sqrt{4} = 2$.

7.21 The posterior density is beta(a+s, b+f) where a = 1, b = 1, s = 2, f = 21. For this density:

$$r = \frac{a+s}{a+s+b+f} = \frac{3}{25} = .120, \quad r^+ = \frac{a+s+1}{a+s+b+f+1} = \frac{4}{26} = .154,$$

$$t = \sqrt{r(r^+ - r)} = .0637, \quad z_{90} = 1.65$$

The 90% posterior probability interval is $.120 \pm 1.65(.0637) = .120 \pm .105$, or from about 1.5% to 22.5%. (Minitab gives slightly different results. The 90% interval has 5% of the probability excluded on both sides. For the beta(3, 22) the fifth percentile is 3.5 and the ninety-fifth percentile is 24. So the exact 90% posterior probability interval is from 3.5% to 24%. The disagreement with the approximation using the Standard Normal Table reflects the lack of symmetry in the beta(3, 22) density—it has a longer right tail than the normal approximation suggests.)

7.23 The posterior density is beta(a+s, b+f) where a = 1, b = 1, s = 181, f = 322. For this density:

$$r = \frac{a+s}{a+s+b+f} = \frac{182}{505} = .3604, \quad r^+ = \frac{a+s+1}{a+s+b+f+1} = \frac{183}{506} = .3662,$$

$$t = \sqrt{r(r^+ - r)} = .02134, \quad z_{95} = 1.96$$

The 95% posterior probability interval is $.360 \pm 1.96 \times .02134$, or from about 31.9% to 40.2%.

7.25 The posterior density is beta(a+s, b+f) where a = 1, b = 9, s = 230, f = 270. For this density:

$$r = \frac{a+s}{a+s+b+f} = \frac{231}{510} = .4529, \quad r^+ = \frac{a+s+1}{a+s+b+f+1} = \frac{232}{511} = .4540,$$

$$t = \sqrt{r(r^+ - r)} = .0220, \quad z_{99} = 2.33$$

The 99% posterior probability interval is $.453 \pm 2.33 \times .0220$, or from about 40.2% to 50.4%.

7.27 For the beta(a, b) density with $a = 2$, $b = 1$:

(a) prob $= \dfrac{a}{a+b} \dfrac{a+1}{a+b+1} \dfrac{a+2}{a+b+2} \dfrac{a+3}{a+b+3} = \dfrac{1}{3}$

(b) prob $= 4 \dfrac{b}{a+b} \dfrac{a}{a+b+1} \dfrac{a+1}{a+b+1} \dfrac{a+2}{a+b+2} = \dfrac{4}{15}$

(c) Probability that the next observation is a success is $\dfrac{a+s}{a+s+b+f} = \dfrac{5}{7}$

7.29 Posterior density: beta(28, 3). The predictive probability that all 10 of them will graduate is

$$\frac{28}{31} \frac{29}{32} \frac{30}{33} \cdots \frac{37}{40} = \frac{28\times29\times30}{38\times39\times40} = .411$$

(The first equality is the result of canceling factors 31, 32, . . ., 37 in both numerator and denominator.)

The predictive probability that 9 of them will graduate is

$$10\times\frac{28}{31} \frac{29}{32} \frac{30}{33} \cdots \frac{36}{39} \frac{3}{40} = \frac{10\times28\times29\times30\times3}{37\times38\times39\times40} = .333$$

The predictive probability that 8 of them will graduate is

$$45\times\frac{28}{31} \frac{29}{32} \frac{30}{33} \cdots \frac{35}{38} \frac{3}{39} \frac{4}{40} = \frac{45\times28\times29\times30\times3\times4}{36\times37\times38\times39\times40} = .167$$

The predictive probability of 8 or 9 or 10 is the sum: $.411 + .333 + .167 = .911$.

CHAPTER 8
COMPARING TWO PROPORTIONS

[For those of you using Minitab, the prior probability tables to exercises 8.7, 8.9, 8.10, 8.11, 8.12 and 8.13 are contained in datafiles. For example, to do exercise 8.11, type the following in Minitab after the prompt MTB >:

```
read 'ex_811' c1-c13
```

This reads the matrix of numbers into columns c1-c13 of the current Minitab worksheet. The values of the first proportion are in c1, the values of the 2nd proportion in c2, and the probabilities start in c3. Then run program 'pp_discm'.]

8.1 (a) The following are the likelihoods divided by the biggest likelihood—the one at $p_W = .8$ and $p_M = .4$, which is why the corresponding entry is 1:

		0	.1	.2	.3	.4	pw .5	.6	.7	.8	.9	1
	0	0	0	0	0	0	0	0	0	0	0	0
	.1	0	0	0	0	0	0	0	.002	.002	0	0
	.2	0	0	0	0	0	.004	.03	.097	.123	.029	0
	.3	0	0	0	0	0	.021	.153	.499	.637	.147	0
	.4	0	0	0	0	0	.033	.241	.785	1.000	.231	0
PM	.5	0	0	0	0	0	.022	.161	.524	.669	.155	0
	.6	0	0	0	0	0	.007	.048	.155	.198	.046	0
	.7	0	0	0	0	0	0	.005	.017	.021	.005	0
	.8	0	0	0	0	0	0	0	0	0	0	0
	.9	0	0	0	0	0	0	0	0	0	0	0
	1	0	0	0	0	0	0	0	0	0	0	0

(b) The posterior probability table is the same as the above table, except that the entries are divided by the sum of the entries in the above table (6.072):

		0	.1	.2	.3	.4	pw .5	.6	.7	.8	.9	1
	0	.0000	.0000	.0000	.0000	.0000	.0000	.0000	.0000	.0000	.0000	.0000
	.1	.0000	.0000	.0000	.0000	.0000	.0000	.0001	.0003	.0003	.0001	.0000
	.2	.0000	.0000	.0000	.0000	.0000	.0007	.0049	.0159	.0203	.0047	.0000
	.3	.0000	.0000	.0000	.0000	.0002	.0035	.0252	.0822	.1048	.0243	.0000
	.4	.0000	.0000	.0000	.0000	.0003	.0054	.0396	.1292	.1647	.0381	.0000
PM	.5	.0000	.0000	.0000	.0000	.0002	.0036	.0265	.0864	.1101	.0255	.0000
	.6	.0000	.0000	.0000	.0000	.0001	.0011	.0078	.0255	.0325	.0075	.0000
	.7	.0000	.0000	.0000	.0000	.0000	.0001	.0009	.0028	.0035	.0008	.0000
	.8	.0000	.0000	.0000	.0000	.0000	.0000	.0000	.0001	.0001	.0000	.0000
	.9	.0000	.0000	.0000	.0000	.0000	.0000	.0000	.0000	.0000	.0000	.0000
	1	.0000	.0000	.0000	.0000	.0000	.0000	.0000	.0000	.0000	.0000	.0000

(This table and other tables given in these answers show the rows in increasing order of the p's. This is consistent with the output from Minitab and flips the order given in the text. The order you use does not matter, except that you have to know which you are using at any given time. For this reason it is essential to label the rows (and columns) with the corresponding population proportions.)

(c) PdAL0 = .9976; PdAL.3 = .7589; PdAL.6 = .0499 (the sum of italicized entries in the table)

8.3 The following are the likelihoods divided by the biggest likelihood (the one at p_C = .6 and p_E = .9):

p_E

		0	.1	.2	.3	.4	.5	.6	.7	.8	.9	1
	0	0	0	0	0	0	0	0	0	0	0	0
	.1	0	0	0	0	0	0	0	0	0	0	0
	.2	0	0	0	0	0	0	0	0	.002	.022	0
	.3	0	0	0	0	0	0	0	0	.011	.147	0
	.4	0	0	0	0	0	0	0	.001	.033	.444	0
p_C	.5	0	0	0	0	0	0	0	.002	.06	.818	0
	.6	0	0	0	0	0	0	0	.003	.074	1.000	0
	.7	0	0	0	0	0	0	0	.002	.059	.798	0
	.8	0	0	0	0	0	0	0	0	.026	.351	0
	.9	0	0	0	0	0	0	0	0	.003	.044	0
	1	0	0	0	0	0	0	0	0	0	0	0

(a) The posterior probability table is the same as the above table, except that the entries are divided by the sum of the entries in the above table:

p_E

		0	.1	.2	.3	.4	.5	.6	.7	.8	.9	1
	0	0	0	0	0	*0*	*0*	*0*	*0*	*0*	*0*	*0*
	.1	0	0	0	0	0	*0*	*0*	*0*	*0*	*0*	*0*
	.2	0	0	0	0	0	0	*0*	*0*	*0*	*.006*	*0*
	.3	0	0	0	0	0	0	0	*0*	*.003*	*.038*	*0*
	.4	0	0	0	0	0	0	0	0	*.008*	*.114*	*0*
p_C	.5	0	0	0	0	0	0	0	.001	.015	*.210*	*0*
	.6	0	0	0	0	0	0	0	.001	.019	.256	*0*
	.7	0	0	0	0	0	0	0	.001	.015	.204	0
	.8	0	0	0	0	0	0	0	0	.007	.090	0
	.9	0	0	0	0	0	0	0	0	.001	.011	0
	1	0	0	0	0	0	0	0	0	0	0	0

(b) PdAL0 = .999, PdAL.2 = .874, PdAL.4 = .378 (the sum of the entries shown in italics in the table), PdAL.6 = .044

8.5 Likelihoods are as follows (divided by the largest likelihood, the one at $p_S = .7$, $p_F = .8$):

		0	.1	.2	.3	.4	.5	.6	.7	.8	.9	1
	0	0	0	0	0	0	0	0	0	0	0	0
	.1	0	0	0	0	0	0	0	0	0	0	0
	.2	0	0	0	0	0	0	0	0	0	0	0
	.3	0	0	0	0	0	0	0	0	0	0	0
	.4	0	0	0	0	0	0	0	0	0	0	0
p_S	.5	0	0	0	0	0	0	.001	.009	.02	.007	0
	.6	0	0	0	0	0	.002	.033	.2	.434	.161	0
	.7	0	0	0	0	0	.005	.076	.46	1.000	.371	0
	.8	0	0	0	0	0	0	.009	.057	.124	.046	0
	.9	0	0	0	0	0	0	0	0	0	0	0
	1	0	0	0	0	0	0	0	0	0	0	0

Column header: PF

Multiplying the above table by the priors and dividing each entry by the sum of the entries gives these posterior probabilities:

		0	.1	.2	.3	.4	.5	.6	.7	.8	.9	1
	0	*0*	0	0	0	0	0	0	0	0	0	0
	.1	0	*0*	0	0	0	0	0	0	0	0	0
	.2	0	0	*0*	0	0	0	0	0	0	0	0
	.3	0	0	0	*0*	0	0	0	0	0	0	0
	.4	0	0	0	0	*0*	0	0	0	0	0	0
p_S	.5	0	0	0	0	0	*0*	0	.002	.003	.001	0
	.6	0	0	0	0	0	0	*.032*	.034	.073	.027	0
	.7	0	0	0	0	0	.001	.013	*.445*	.168	.062	0
	.8	0	0	0	0	0	0	.002	.010	*.120*	.008	0
	.9	0	0	0	0	0	0	0	0	0	*0*	0
	1.0	0	0	0	0	0	0	0	0	0	0	*0*

Column header: PF

The total probability on the diagonal (italicized in the table) is .599. This is the probability that Carr has the same chance of making a shot after a success as after a failure.

8.7 **(a)** The tables on the next page are both correct. The first is in the scheme of the prior table given in the text and the second gives the rows in the opposite order (as given by Minitab). Both show the entries on the diagonal $p_M = p_W$ in italics. Their sum is the answer: $P(p_M = p_F) = .002 + .020 + .042 + .015 = .079$.

(b) PdAL0 = .002+.02+.042+.015+.001+.001+.001 = .082
PdAL.1 = .001+.001+.001 = .003
The difference is .082−.003 = .079, as in part (a).

Posterior probabilities for Exercise 8.7: Model for Women p_W vs. Model for Men p_M										
p_W\p_M	.1	.2	.3	.4	.5	.6	.7	.8	.9	sum
.9	.000	.000	.000	.000	.000	.000	.000	.000	*.000*	.000
.8	.000	.000	.000	.000	.000	.000	.001	*.000*	.000	.001
.7	.000	.000	.000	.000	.000	.001	*.015*	.003	.001	.020
.6	.000	.000	.000	.000	.001	*.042*	.024	.030	.007	.104
.5	.000	.000	.000	.000	*.020*	.025	.081	.103	.024	.253
.4	.000	.000	.000	*.002*	.005	.037	.121	.154	.036	.355
.3	.000	.000	*.000*	.000	.003	.024	.077	.098	.023	.225
.2	.000	*.000*	.000	.000	.001	.005	.015	.019	.004	.044
.1	*.000*	.000	.000	.000	.000	.000	.000	.000	.000	.000
sum	.000	.000	.000	.002	.030	.134	.334	.407	.095	1

Posterior probabilities for Exercise 8.7: Model for Women p_W vs. Model for Men p_M										
p_W\p_M	.1	.2	.3	.4	.5	.6	.7	.8	.9	sum
.1	*.000*	.000	.000	.000	.000	.000	.000	.000	.000	.000
.2	.000	*.000*	.000	.000	.001	.005	.015	.019	.004	.044
.3	.000	.000	*.000*	.000	.003	.024	.077	.098	.023	.225
.4	.000	.000	.000	*.002*	.005	.037	.121	.154	.036	.355
.5	.000	.000	.000	.000	*.020*	.025	.081	.103	.024	.253
.6	.000	.000	.000	.000	.001	*.042*	.024	.030	.007	.104
.7	.000	.000	.000	.000	.000	.001	*.015*	.003	.001	.020
.8	.000	.000	.000	.000	.000	.000	.001	*.000*	.000	.001
.9	.000	.000	.000	.000	.000	.000	.000	.000	*.000*	.000
sum	.000	.000	.000	.002	.030	.134	.334	.407	.095	1

8.9 The posterior probability table is as follows:

p_H\p_S	.4	.5	.6	.7	.8	.9	sum
.4	.0000	.0000	.0000	.0000	.0000	.0000	.0000
.5	.0000	.0000	.0000	.0000	.0000	.0000	.0000
.6	.0000	.0003	.0011	.0004	.0000	.0000	.0018
.7	.0005	.0101	.0271	.0171	.0005	.0000	.0553
.8	.0044	.0635	.2559	.1210	.0069	.0000	.4517
.9	.0030	.0861	.2311	.1639	.0070	.0000	.4911
sum	.0089	.1600	.5152	.3024	.0144	.0000	1.0000

(The sums showing the probabilities of p_H and p_S separately are extra and are not necessary to complete this exercise.) The posterior probability of $p_H > p_S$ is PdAL.1 where $d = p_H - p_S$. This is the sum of the entries in the table below the diagonal: .9749.

8.11 These are the posterior probabilities:

PI\PN	0	.1	.2	.3	.4	.5	.6	.7	.8	.9	1
0	*.0000*	.0000	.0000	.0000	.0000	.0000	.0000	.0000	.0000	.0000	.0000
.1	.0001	*.0014*	.0000	.0000	.0000	.0000	.0000	.0000	.0000	.0000	.0000
.2	.0011	.0007	*.0062*	.0002	.0001	.0000	.0000	.0000	.0000	.0000	.0000
.3	.0039	.0023	.0013	*.0108*	.0003	.0001	.0000	.0000	.0000	.0000	.0000
.4	.0092	.0054	.0030	.0015	*.0119*	.0003	.0001	.0000	.0000	.0000	.0000
.5	.0179	.0106	.0059	.0030	.0014	*.0093*	.0002	.0000	.0000	.0000	.0000
.6	.0309	.0182	.0101	.0052	.0024	.0010	*.0053*	.0001	.0000	.0000	.0000
.7	.0491	.0290	.0161	.0082	.0038	.0015	.0005	*.0020*	.0000	.0000	.0000
.8	.0732	.0432	.0240	.0123	.0057	.0023	.0008	.0002	*.0004*	.0000	.0000
.9	.1043	.0616	.0342	.0175	.0081	.0033	.0011	.0003	.0000	*.0000*	.0000
1	.1430	.0845	.0469	.0240	.0111	.0045	.0015	.0003	.0000	.0000	*.0000*

(a) The posterior probability of $d = 0$ is the sum of the (italicized) entries with $p_I = p_N$: .047. (This is down from .67, suggesting that the evidence in favor of a difference is pretty strong.)

(b) The posterior probability of $d > 0$ is PdAL.1; this is the sum of the entries in the table below the diagonal with italicized entries: .951.

8.13 **(a)** The following table gives my posterior probabilities:

						p_B				
		.1	.2	.3	.4	.5	.6	.7	.8	.9
	.1	.0113	.0440	.1496	.2255	.1136	.0198	.0009	.0000	.0000
	.2	.0000	.2437	.0442	.0333	.0084	.0015	.0001	.0000	.0000
	.3	.0000	.0002	.0986	.0024	.0006	.0001	.0000	.0000	.0000
	.4	.0000	.0000	.0000	.0021	.0000	.0000	.0000	.0000	.0000
p_I	.5	.0000	.0000	.0000	.0000	.0000	.0000	.0000	.0000	.0000
	.6	.0000	.0000	.0000	.0000	.0000	.0000	.0000	.0000	.0000
	.7	.0000	.0000	.0000	.0000	.0000	.0000	.0000	.0000	.0000
	.8	.0000	.0000	.0000	.0000	.0000	.0000	.0000	.0000	.0000
	.9	.0000	.0000	.0000	.0000	.0000	.0000	.0000	.0000	.0000

(b) My posterior probability of the null hypothesis $p_I = p_B$ is the sum of the entries on the diagonal: .0113+.2437+.0986+.0021 = .3557.

(c) PdAL.1 = .6440, which is essentially one minus that answer in (b) because very little posterior probability is associated with $d < 0$ (so I am pretty convinced that Insight therapy is not harmful to a marriage, although I remain skeptical that it has a benefit); PdAL.2 = .5534.

31

CHAPTER 9
DENSITIES FOR TWO PROPORTIONS

9.1 (**a**) For Ted Williams, the posterior density is beta(105+2654, 245+5052) = beta(2759, 5297). So

$$r_W = \frac{2759}{2759+5297} = .342478, \quad r_W^+ = \frac{2759+1}{2759+5297+1} = .342559,$$

$$t_W^2 = r_W(r_W^+ - r_W) = 2.793E\text{-}5$$

Similarly, for Joe Dimaggio, the posterior density is beta(105+2214, 245+6821-2214) = beta(2319, 4852). So $r_D = .323386$, $r_D^+ = .323480$, and $t_D^2 = 3.051E\text{-}5$. Thus

$$z = \frac{0 - (r_W - r_D)}{\sqrt{t_W^2 + t_D^2}} = -2.50$$

PdAL0 is the probability to the right of -2.50, which from the Standard Normal Table is 1 - .0062 = .9938.

(**b**) For Williams, the posterior density is beta(540, 7645). So $r_W = .0659743$ and $t_W^2 = 7.528E\text{-}6$.

For Dimaggio, the posterior density is beta(380, 6920). So $r_D = .0520548$, $t_D^2 = 6.759E\text{-}6$.

The z-score for 0 is z = -3.68. The probability to the right of z = -3.68 is .9999.

[If you use the Minitab program 'pp_beta' for these calculations you will get slightly different answers. One reason for this is that the normal curve and the Standard Normal Table involves an approximation while Minitab does not. However, Minitab uses simulation and this means that you'll get different answers each time you use it. I used it twice (1000 simulations for each) for (a) getting PdAL0 = .996 and .994 (as opposed to .9938 above) and twice for (b) getting 1.000 both times (so in none of the total of 2000 simulations did it happen that $p_W < p_D$).]

9.3   (**a**) The posterior densities for p_E and p_C are beta(1+28, 1+1) = beta(29, 2) and beta(1+6, 1+4) = beta(7, 5).

    (**b**) r_E = .93548, r_E^+ = .93750, t_E = .04343, r_C = .58333, r_C^+ = .61538, t_C = .13674

x	0	.2	.4	.6
z-score	-2.45	-1.06	.33	1.73
PdALx	.9927	.8554	.3707	.0418

[One can also calculate these PdAL's using Minitab program 'pp_beta'. For two different runs of 1000 simulations each I got these values:

	x	0	.2	.4	.6
Run 1:	PdALx	.996	.838	.374	.047
Run 2:	PdALx	.996	.848	.342	.037]

9.5   Using beta(1, 1) for both death rates, and thereby greatly discounting the prior information, gives these posterior densities:

     p_T is beta(1+1, 1+144) = beta(2, 145)

     p_C is beta(1+16, 1+121) = beta(17, 122)

(Your posterior densities will of course be different from these if your priors are not both beta(1, 1).) The following are needed to calculate z:

$$r_T = \frac{2}{147} = .0136054, \quad r_T^+ = \frac{3}{148} = .0202703, \quad t_T = \sqrt{r_T\,(r_T^+ - r_T)} = .00952$$

$$r_C = \frac{17}{139} = .1223022, \quad r_C^+ = \frac{18}{140} = .1285714, \quad t_C = \sqrt{r_C\,(r_C^+ - r_C)} = .02769$$

Let $d = p_C - p_T$, so a positive d means AZT is an improvement. The z-score for d = 0 is

$$z = \frac{0 - (r_C - r_T)}{\sqrt{t_C^2 + t_T^2}} = -3.71$$

and PdAL0 = .9999. For d = .05, a five-percentage-point improvement,

$$z = \frac{.05 - (r_C - r_T)}{\sqrt{t_C^2 + t_T^2}} = -2.00$$

and PdAL.05 = .9772. For d = .1, a ten-percentage-point improvement,

$$z = \frac{.1 - (r_C - r_T)}{\sqrt{t_C^2 + t_T^2}} = -.30$$

and PdAL.1 = .6179. [Using Minitab program 'pp_beta' with 1000 simulations for these calculations I got PdAL0 = 1.000, PdAL.05 = .985 and PdAL.1 = .604. A second set of 1000 simulations gave 1.000, .989 and .608.]

9.7 **(a)** Repeating from 9.1, $r_W = .342478$, $r_W^+ = .342559$, $t_W^2 = 2.793E\text{-}5$

Similarly, $r_D = .323386$, $r_D^+ = .323480$, and $t_D^2 = 3.051E\text{-}5$

Thus

$$r_W - r_D \pm z_{\text{perc}} \sqrt{t_W^2 + t_D^2} = .0191 \pm 1.96\sqrt{5.844E\text{-}5} = .0191 \pm .0150$$

That is, the 95% posterior probability interval extends from .0041 to .0341.

(b) Since the null hypothesis value of 0 difference is not contained in the interval in part (a), this hypothesis is not supported. There is still some uncertainty concerning whether Williams was a better hitter than DiMaggio, but the evidence is at least moderately convincing that he was.

9.9 $s_T = 13$, $f_T = 144$, $s_C = 40$, $f_C = 117$, $z_{95} = 1.96$

(a) $a_T = a_C = 1$, $b_T = b_C = 1$

$r_T = .08805$, $r_T^+ = .09375$, $t_T^2 = .0005019$

$r_C = .25786$, $r_C^+ = .26250$, $t_C^2 = .001196$

$$r_T - r_C \pm z_{\text{perc}} \sqrt{t_T^2 + t_C^2} = -.170 \pm .081$$

The 95% posterior probability interval is from -25.1% to -8.9%.

(b) $a_T = b_T = 1$, $a_C = 25$, $b_C = 75$

$r_T = .08805$, $r_T^+ = .09375$, $t_T^2 = .0005019$

$r_C = .25292$, $r_C^+ = .25581$, $t_C^2 = .0007324$

$$r_T - r_C \pm z_{\text{perc}} \sqrt{t_T^2 + t_C^2} = -.165 \pm .069$$

The 95% posterior probability interval is from -23.4% to -9.6%.

(c) $a_T = 2$, $b_T = 18$, $a_C = 25$, $b_C = 75$

$r_T = .08475$, $r_T^+ = .08989$, $t_T^2 = .0004357$

$r_C = .25292$, $r_C^+ = .25581$, $t_C^2 = .0007324$

$$r_T - r_C \pm z_{\text{perc}} \sqrt{t_T^2 + t_C^2} = -.168 \pm .067$$

The 95% posterior probability interval is from -23.5% to -10.1%.

The three intervals in parts (a), (b), and (c) are different even though all three refer to the same data. The reason is that the prior densities are different. But the intervals are not *very* different because the data dominate in the posterior density. The conclusions are not very sensitive to the prior density and so people with moderately different prior opinions will come to roughly the same conclusions—the conclusions are **robust** with respect to the prior.

9.11 The posterior density of p_A is beta(1+28, 1+21) = beta(29, 22) and that of p_N is beta(1+1, 1+8) = beta(2, 9).

$$r_A = .5686, \; r_A{}^+ = .5769, \; t_A = .06868$$

$$r_N = .1818, \; r_N{}^+ = .2500, \; t_N = .11134$$

(a) $r_A - r_N \pm z_{perc} \sqrt{t_A^2 + t_N^2} = .387 \pm 1.65(.1308) = .387 \pm .216$

That is, the 90% posterior probability interval for $d = p_A - p_N$ extends from 17.1% to 60.3%.

(b) The 95% posterior interval is $.387 \pm .256$; that is, from 13.0% to 64.3%. The null hypothesis value $d = 0$ is not contained in this interval and so it is not supported. The evidence strongly suggests that an abstract's conclusion about the deleterious effects of cocaine affects its acceptability to peers.

9.13 The posterior density of p_T is beta(5+20, 5+10) = beta(25, 15) and that of p_C is beta(5+9, 5+22) = beta(14, 27).

$$r_T = .62500, \; r_T{}^+ = .63415, \; t_T = .075607$$

$$r_C = .34146, \; r_C{}^+ = .35714, \; t_C = .073171$$

Calculations shown below use the following formula and the Standard Normal Table:

$$z = \frac{x - (r_T - r_C)}{\sqrt{t_T^2 + t_C^2}}$$

x	0	.1	.2	.3	.4	.5	.6	.7	.8	.9	1
z	-2.69	-1.74	-.79	.16	1.11	2.06	3.01	3.96	4.91	5.86	6.81
PdALx	.996	.959	.785	.436	.133	.020	.001	.000	.000	.000	.000

So there is a large probability (over 99%) that the drug has a beneficial effect in the sense that more patients walk further on drug than on placebo. And there is a reasonable chance (about 44%) that 30% or more of the patients walk further on drug than on placebo.

9.15 The posterior density of p_U is beta(1+16, 2+20) = beta(17, 22) and that of p_N is beta(1+14, 2+25) = beta(15, 27).

$r_U = .43590$, $r_U^+ = .45000$, $t_U = .078405$

$r_N = .35714$, $r_N^+ = .37209$, $t_N = .073071$

(a) Calculations shown below use the following formula and the Standard Normal Table:

$$z = \frac{x - (r_U - r_N)}{\sqrt{t_U^2 + t_N^2}}$$

x	0	.1	.2
z	-.73	.20	1.13
PdALx	.767	.421	.129

Based on prior information the value of PdAL0 was .50. It has increased somewhat because a greater proportion of Underweight women in the sample subscribed. But it has not increased much because the evidence is not very strong.

(b) $r_U - r_N \pm z_{perc} \sqrt{t_U^2 + t_N^2} = .079 \pm 1.65(.10718) = .079 \pm .177$

That is, the 90% posterior probability interval extends from -9.8% to 25.6%.

(c) The 95% posterior probability interval extends from -13.1% to 28.9%. This interval does not contain 0 and so the null hypothesis d = 0 is supported.

9.17 The posterior density of p_P is beta(1+11, 3+27) = beta(12, 30) and that of p_N is beta(1+2, 3+26) = beta(3, 29).

$r_P = .28571$, $r_P^+ = .30233$, $t_P = .06889$

$r_N = .09375$, $r_N^+ = .12121$, $t_N = .05074$

(a) $r_P - r_N \pm z_{perc} \sqrt{t_P^2 + t_N^2} = .192 \pm 1.28(.08556) = .192 \pm .110$

That is, the 80% posterior probability interval extends from 8.2% to 30.2%.

(b) The null hypothesis that there is no difference in rates of sexual abuse as a child is not supported because d = 0 is not contained in the 95% posterior probability interval, which is $.192 \pm 1.96(.08556)$, extending from 2.4% to 36.0%. The evidence in favor of a relationship is rather strong, but it is not conclusive; in particular, a 99% probability interval does not contain 0.

9.19 The posterior density of p_M is beta(1+8, 1+12) = beta(9, 13) and that of p_W is beta(1+17, 1+5) = beta(18, 6).

$r_M = .40909$, $r_M{}^+ = .43478$, $t_M = .1025$

$r_W = .75000$, $r_W{}^+ = .76000$, $t_W = .0866$

(a) This is PdAL0. As given in part (b), PdAL0 = .994.

(b) Calculations shown below use the following formula and the Standard Normal Table:

$$z = \frac{x - (r_W - r_M)}{\sqrt{t_W^2 + t_M^2}}$$

x	0	.3	.6
z	-2.54	-.30	1.93
PdALx	.994	.618	.027

(c) $r_W - r_M \pm z_{perc} \sqrt{t_W^2 + t_M^2} = .341 \pm 2.58(.1342) = .341 \pm .346$

That is, the 99% posterior probability interval for d extends from -.5% to 68.7%.

(d) The 95% posterior probability interval for d extends from 7.8% to 60.4%. This interval does not contain 0 and so the null hypothesis is not supported. There is rather clear evidence that men are less likely to think that there is too much violence on television (at least in the limited setting of this survey), but the paucity of data makes the 95% probability interval for d rather broad.

9.21 (a) The posterior density of p_G is beta(9+64, 1+18) = beta(73, 19). The posterior density of p_C is beta(5+53, 5+25) = beta(58, 30).

$r_G = .79348$, $r_G{}^+ = .79570$, $t_G = .04198$

$r_C = .65909$, $r_C{}^+ = .66292$, $t_C = .05024$

$$r_G - r_C \pm z_{perc} \sqrt{t_G^2 + t_C^2} = .134 \pm 1.96(.06547) = .134 \pm .128$$

That is, the 95% posterior probability interval for d extends from .6% to 26.3%. This interval does not contain 0 and so, when combined with this prior information, the evidence does not support equality of the two treatments at 6 weeks.

(b) The posterior density of p_G is beta(9+39, 1+32) = beta(48, 33). The posterior density of p_C is beta(5+43, 5+27) = beta(48, 32).

$$r_G = .59259, \quad r_G^+ = .59756, \quad t_G = .05426$$

$$r_C = .60000, \quad r_C^+ = .60494, \quad t_C = .05443$$

$$r_G - r_C \pm z_{perc} \sqrt{t_G^2 + t_C^2} = -.007 \pm 1.96(.07686) = -.007 \pm .151$$

That is, the 95% posterior probability interval for d extends from -15.8% to 14.3%. This interval contains 0 and so the evidence supports equality of the performance of the two treatments at 24 weeks. Actually, now the center of this interval is slightly to the left of 0 since the evidence at 24 weeks actually points to the control being better, even when including the prior information.

9.23 The posterior density of p_F is beta(1+267, 1+78) = beta(268, 79) and that of p_S is beta(1+250, 1+95) = beta(91, 166).

$$r_F = .77233, \quad r_F^+ = .77299, \quad t_F = .02248$$

$$r_S = .72334, \quad r_S^+ = .72414, \quad t_S = .02398$$

(a) $\quad r_F - r_S \pm z_{perc} \sqrt{t_F^2 + t_S^2} = .049 \pm 1.65(.03287) = .049 \pm .054$

That is, the 90% posterior probability interval for d extends from -.5% to 10.3%.

(b) Calculations shown below use the following formula and the Standard Normal Table:

$$z = \frac{x - (r_F - r_S)}{\sqrt{t_F^2 + t_S^2}}$$

x	0	.05	.1
z	-1.49	.03	1.55
PdALx	.932	.488	.061

9.25 The posterior density of p_F is beta(1+8, 1+20) = beta(9, 21) and that of p_M is beta(1+9, 1+19) = beta(10, 20).

$$r_F = .30000, \quad r_F^+ = .32258, \quad t_F = .08230$$

$$r_M = .33333, \quad r_M^+ = .35484, \quad t_M = .08467$$

$$r_F - r_M \pm z_{perc} \sqrt{t_F^2 + t_M^2} = -.033 \pm 1.96(.11808) = -.0333 \pm .2314$$

That is, the 95% posterior probability interval for d extends from about -26.5% to 19.8%.

9.27 (a) The posterior density of p_D is beta(1+122, 1+80) = beta(123, 81) and that of p_C is beta(1+86, 1+113) = beta(87, 114).

$r_D = .60294$, $r_D^+ = .60488$, $t_D = .03417$

$r_C = .43284$, $r_C^+ = .43564$, $t_C = .03486$

Calculations shown below use the following formula and the Standard Normal Table:

$$z = \frac{x - (r_D - r_C)}{\sqrt{t_D^2 + t_C^2}}$$

x	0	.1	.2
z	-3.48	-1.44	.61
PdALx	1.000	.925	.271

(b) The posterior density of p_E is beta(1+93, 1+109) = beta(94, 110) and that of p_C is beta(1+86, 1+113) = beta(87, 114).

$r_E = .46078$, $r_E^+ = .46341$, $t_E = .03481$

$r_C = .43284$, $r_C^+ = .43564$, $t_C = .03486$

Calculations shown below use the following formula and the Standard Normal Table:

$$z = \frac{x - (r_E - r_C)}{\sqrt{t_E^2 + t_C^2}}$$

x	0	.1	.2
z	-.57	1.46	3.49
PdALx	.716	.072	.000

(c) The posterior density of p_B is beta(1+123, 1+74) = beta(124, 75) and that of p_E is beta(1+93, 1+109) = beta(94, 110).

$r_B = .62312$, $r_B^+ = .62500$, $t_B = .03427$

$r_E = .46078$, $r_E^+ = .46341$, $t_E = .03481$

The table below uses the following formula and the Standard Normal Table:

$$z = \frac{x - (r_B - r_E)}{\sqrt{t_B^2 + t_E^2}}$$

x	0	.1	.2
z	-3.32	-1.28	.77
PdALx	1.000	.900	.221

Summarizing the three parts of this exercise, for these prior assumptions it is pretty clear that Deprenyl is better than placebo and that Both deprenyl and vitamin E is better than vitamin E alone. However, it is not clear whether vitamin E has a benefit when used alone.

9.29 Let p_A be the rate of conversion for the population who Always used condoms and p_N be the corresponding rate for those who did Not always use them. The posterior density of p_A is beta(1+3, 1+168) = beta(4, 169) and that of p_N is beta(1+16, 1+118) = beta(17, 119).

$$r_A = .02312, \; r_A^+ = .02874, \; t_A = .01139$$

$$r_N = .12500, \; r_N^+ = .13139, \; t_N = .02825$$

The 95% posterior probability interval is as follows:

$$r_N - r_A \pm z_{perc} \sqrt{t_N^2 + t_A^2} = .102 \pm 1.96(.03047) = .102 \pm .060$$

That is, the 95% posterior probability interval extends from 4.2% to 16.2%. The null hypothesis that there is no difference in rates of HIV-Sero conversion is not supported because $d = 0$ is not contained in this interval.

9.31 Let p_S be the rate of seropositive conversion for the Spermicide population and p_C be the corresponding rate for placebo (Control). The posterior density of p_S is beta(1+27, 1+33) = beta(28, 34) and that of p_C is beta(1+20, 1+36) = beta(21, 37).

$$r_S = .45161, \; r_S^+ = .46032, \; t_S = .06270$$

$$r_C = .36207, \; r_C^+ = .37288, \; t_C = .06257$$

The 95% posterior probability interval is as follows:

$$r_S - r_C \pm z_{perc} \sqrt{t_S^2 + t_C^2} = .090 \pm 1.96(.08858) = .090 \pm .174$$

That is, the 95% posterior probability interval extends from -8.4% to 26.3%. The null hypothesis that there is no benefit to using this spermicide is supported because $d = 0$ is contained in this interval. The fact that Control actually performed better makes even clearer the lack of a benefit for Spermicide.

9.33 **(a)** Let p_1 be the rate of defectives for lot **1** and p_2 be the corresponding rate for lot **2**. The posterior density of p_1 is beta(1+11, 1+29) = beta(12, 30) and that of p_2 is beta(1+11, 1+9) = beta(12, 10).

$r_1 = .28571$, $r_1^+ = .30233$, $t_1 = .06889$

$r_2 = .54545$, $r_2^+ = .56522$, $t_2 = .10382$

The 95% posterior probability interval is as follows:

$$r_1 - r_2 \pm z_{perc} \sqrt{t_1^2 + t_2^2} = -.260 \pm 1.96(.1246) = -.260 \pm .244$$

That is, the 95% posterior probability interval extends from -50.4% to -1.6%. The null hypothesis that there is no difference in defective rates in these two lots is not supported because d = 0 is not contained in this interval. The available evidence suggests a real difference in the lots' defective rates.

(b) Let p_3 be the rate of defectives for lot **3**. The posterior density of p_1 is as in (a) and that of p_3 is beta(1+23, 1+5) = beta(24, 6).

$r_1 = .28571$, $r_1^+ = .30233$, $t_1 = .06889$

$r_3 = .80000$, $r_3^+ = .80645$, $t_3 = .07184$

The 95% posterior probability interval is as follows:

$$r_1 - r_3 \pm z_{perc} \sqrt{t_1^2 + t_3^2} = -.514 \pm 1.96(.09954) = -.514 \pm .195$$

That is, the 95% posterior probability interval extends from -70.9% to -31.9%. The null hypothesis that there is no difference in defective rates in these two lots is not supported because d = 0 is not contained in this interval. The available evidence suggests a real difference in the lots' defective rates.

(c) Let p_4 be the rate of defectives for lot **4**. The posterior density of p_1 is as in (a) and that of p_4 is beta(1+14, 1+50) = beta(15, 51).

$r_1 = .28571$, $r_1^+ = .30233$, $t_1 = .06889$

$r_4 = .22727$, $r_4^+ = .23881$, $t_4 = .05120$

The 95% posterior probability interval is as follows:

$$r_1 - r_4 \pm z_{perc} \sqrt{t_1^2 + t_4^2} = -.058 \pm 1.96(.08583) = -.058 \pm .168$$

That is, the 95% posterior probability interval extends from -11.0% to 22.7%. The null hypothesis that there is no difference in defective rates in these two lots is supported because d = 0 is contained in this interval. The available evidence does not suggest a real difference in the lots' defective rates.

(**d**) Let p_5 be the rate of defectives for lot **5**. The posterior density of p_1 is as in (a) and that of p_5 is beta(1+1, 1+15) = beta(2, 16).

$r_1 = .28571$, $r_1^+ = .30233$, $t_1 = .06889$

$r_5 = .11111$, $r_5^+ = .15789$, $t_5 = .07210$

The 95% posterior probability interval is as follows:

$$r_1 - r_5 \pm z_{perc} \sqrt{t_1^2 + t_5^2} = .175 \pm 1.96(.09972) = .175 \pm .195$$

That is, the 95% posterior probability interval extends from -2.1% to 37.0%. The null hypothesis that there is no difference in defective rates in these two lots is supported because d = 0 is contained in this interval. The available evidence does not suggest a real difference in the lots' defective rates.

Summarizing the various parts of this exercise, the null hypothesis that lots 1 and 4 and lots 1 and 5 have the same rate of defectives is supported, but lots 1 and 2 and lots 1 and 3 seem to have different rates.

9.35 Let p_H be the rate of small business ownership among men who were **H**yperactive as boys and p_N the corresponding rate among men who were **N**ot hyperactive as boys. Let $d = p_H - p_N$. The posterior density of p_H is beta(1+16, 5+75) = beta(17, 80) and that of p_N is beta(1+5, 5+90) = beta(6, 95).

$r_H = .17526$, $r_H^+ = .18367$, $t_H = .03840$

$r_N = .05941$, $r_N^+ = .06863$, $t_N = .02340$

The 95% posterior probability interval is as follows:

$$r_H - r_N \pm z_{perc} \sqrt{t_H^2 + t_N^2} = .116 \pm 1.96(.04497) = .116 \pm .088$$

So the 95% posterior probability interval extends from 2.8% to 20.4%. The null hypothesis that there is no difference in rates of small business ownership is not supported because d = 0 is not contained in this interval—the observed difference is probably real.

9.37 Let p_I be the population proportion of **I**dentical twinships in which both members are alcoholics, and p_F the corresponding proportion of **F**raternal twinships. Let $d = p_I - p_F$. The posterior density of p_I is beta(1+8, 1+45) = beta(9, 46) and that of p_F is beta(1+4, 1+59) = beta(5, 60).

$r_I = .16364$, $r_I^+ = .17857$, $t_I = .04944$

$r_F = .07692$, $r_F^+ = .09091$, $t_F = .03280$

The 95% posterior probability interval is as follows:

$$r_I - r_F \pm z_{perc} \sqrt{t_I^2 + t_F^2} = .087 \pm 1.96(.05933) = .087 \pm .116$$

So the 95% posterior probability interval extends from -3.0% to 20.3%. The null hypothesis that there is no difference in population proportions of "both" is supported because d = 0 is contained in this interval. From evidence it is not clear that genetic factors play *any* role in alcoholism, far less a "major etiologic role".

9.39 **(a)** Let p_I be the population proportion of **I**dentical twinships in which the cotwin is Lesbian, and p_F the corresponding proportion of **F**raternal twinships. Let d = p_I - p_F. The posterior density of p_I is beta(1+27, 1+29) = beta(28, 30) and that of p_F is beta(1+14, 1+43) = beta(15, 44).

r_I = .48276, r_I^+ = .49153, t_I = .06506

r_F = .25424, r_F^+ = .26667, t_F = .05621

The 95% posterior probability interval is as follows:

$$r_I - r_F \pm z_{perc} \sqrt{t_I^2 + t_F^2} = .229 \pm 1.96(.08598) = .229 \pm .169$$

So the 95% posterior probability interval extends from 6.0% to 39.7%. As in the previous exercise, the null hypothesis that there is no difference in population proportion in which the cotwin is Lesbian is not supported because d = 0 is not contained in this interval. Now the interval is closer to 0, but the two experiments in the different sexes taken together enhance the strength of the overall conclusion.

(b) Let p_A be the population proportion of **A**doptive sisters who are Lesbian. Let d = p_F - p_A. The posterior density of p_F is as in (a) and that of p_A is beta(1+5, 1+27) = beta(6, 28).

r_F = .25424, r_F^+ = .26667, t_F = .05621

r_A = .17647, r_A^+ = .20000, t_A = .06444

The 95% posterior probability interval is as follows:

$$r_F - r_A \pm z_{perc} \sqrt{t_F^2 + t_A^2} = .078 \pm 1.96(.08551) = .078 \pm .168$$

So the 95% posterior probability interval extends from -9.0% to 24.5%. The null hypothesis that there is no difference in population proportion in which the brother is Lesbian is not supported because d = 0 is not contained in this interval. (The conclusion is different from the one in (a) in part because of the small sample size in the Adoptive sisters group.)

CHAPTER 10
GENERAL SAMPLES AND POPULATION MEANS

[In some solutions I will use Minitab and in some I will not; when I use Minitab I will sometimes give a detailed solution as well.]

10.1 In Example 10.1 there were 4 ones, no twos, 10 threes, 9 fours, 6 fives and 1 six. As in that example:

$$P(D|M1) = \left(\tfrac{3}{18}\right)^{30} = 3^{30}\left(\tfrac{1}{18}\right)^{30}$$

$$P(D|M2) = \left(\tfrac{1}{18}\right)^{5}\left(\tfrac{3}{18}\right)^{6}\left(\tfrac{5}{18}\right)^{19} = 1^{5}\,3^{6}\,5^{19}\left(\tfrac{1}{18}\right)^{30}$$

$$P(D|M3) = \left(\tfrac{1}{18}\right)^{4}\left(\tfrac{3}{18}\right)^{19}\left(\tfrac{5}{18}\right)^{7} = 1^{4}\,3^{19}\,5^{7}\left(\tfrac{1}{18}\right)^{30}$$

Now, for the additional model M4:

$$P(D|M4) = \left(\tfrac{57}{396}\right)^{4}\left(\tfrac{29}{396}\right)^{0}\left(\tfrac{142}{396}\right)^{10}\left(\tfrac{114}{396}\right)^{9}\left(\tfrac{27}{396}\right)^{6}\left(\tfrac{27}{396}\right)^{1}$$

$$= \left(\tfrac{57}{22}\right)^{4}\left(\tfrac{142}{22}\right)^{10}\left(\tfrac{114}{22}\right)^{9}\left(\tfrac{27}{22}\right)^{7}\left(\tfrac{1}{18}\right)^{30}$$

As in the example, the factor $(1/18)^{30}$ appears in all likelihoods and so we can drop it throughout:

Likelihood of M1: $3^{30} = 21\text{ E+13}$

Likelihood of M2: $1^{5}\,3^{6}\,5^{19} = 1390\text{ E+13}$

Likelihood of M3: $1^{4}\,3^{19}\,5^{7} = 9\text{ E+13}$

Likelihood of M4: $\left(\tfrac{57}{22}\right)^{4}\left(\tfrac{142}{22}\right)^{10}\left(\tfrac{114}{22}\right)^{9}\left(\tfrac{27}{22}\right)^{7} = 6388\text{ E+13}$

Again, all that matters are likelihoods in relation to the other likelihoods, so we can also drop the E+13 from all four. The posterior probabilities are found as follows, and are given in the last column of the table on the next page:

Model	Likeli for D	P(model)	Product	P(model|D)
M1	21	1/4	21/4	21/7808 = .003
M2	1390	1/4	1390/4	1390/7808 = .178
M3	9	1/4	9/4	9/7808 = .001
M4	6388	1/4	6388/4	6388/7808 = .818
Sums		1	7808/4	1.000

10.3 (a) I carried out the experiment in question and got 0 heads 10 times, 1 head 26 times, and 2 heads 14 times. Your results will most likely be different from mine and so too will your answer to the next two parts of this exercise.

(b) The likelihoods are as follows:

$$P(data|Jon) = \left(\frac{1}{3}\right)^{10}\left(\frac{1}{3}\right)^{26}\left(\frac{1}{3}\right)^{14} = \left(\frac{1}{3}\right)^{50}$$

$$P(data|Sue) = \left(\frac{1}{4}\right)^{10}\left(\frac{2}{4}\right)^{26}\left(\frac{1}{4}\right)^{14} = 2^{26}\left(\frac{1}{4}\right)^{50}$$

Using Bayes' rule the posterior probability that Jon is correct is

$$P(Jon|data) = \frac{P(data|Jon)P(Jon)}{P(data|Jon)P(Jon) + P(data|Sue)P(Sue)}$$

$$= \frac{\left(\frac{1}{3}\right)^{50}\frac{1}{2}}{\left(\frac{1}{3}\right)^{50}\frac{1}{2} + 2^{26}\left(\frac{1}{4}\right)^{50}\frac{1}{2}} = .0256$$

And 1 minus this gives the posterior probability that Sue is correct:

P(Sue|data) = 1 - .0256 = .9744

(c) Using the law of total probability, the predictive probability of 1 head (and 1 tail) in the next toss of a pair of coins is

$$.0256\,\frac{1}{3} + .9744\,\frac{1}{2} = .4957$$

This is very nearly that of Sue's model because the data give much support to her model.

10.5 This is the Minitab input and output:

```
MTB > name c1 'm' c2 'prior'
MTB > set 'm'
DATA> 0 3
DATA> end
MTB > set 'prior'
DATA> .5 .5
DATA> end
MTB > exec 'm_disc'

INPUT POPULATION STANDARD DEVIATION:
DATA> 5

OBSERVED DATA IN WORKSHEET? (TYPE 'y' OR 'n'.)
  IF YES, INPUT NUMBER OF COLUMN.
  IF NO, INPUT OBSERVED SAMPLE MEAN AND SAMPLE SIZE:
n
DATA> 2.406 32

MEAN
  2.406

COUNT
   32

ROW    m    prior   M_x_PRIO   LIKE    PRODUCT     POST    M_x_POST

 1     0    0.5       0.0      30837     15419   0.029915   0.00000
 2     3    0.5       1.5    1000000    500000   0.970086   2.91026
 3                    1.5                                   2.91026
```

So the posterior probability of m = 0 is about 3% and that of m = 3 is about 97%.

(The calculation of likelihoods is as follows:

For m = 0 the z-score is (carrying lots of decimal accuracy)

$$z = \sqrt{n}\, \frac{\bar{x} - m}{h} = \sqrt{32}\, \frac{2.406 - 0}{5} = 2.7220783$$

And

$$e^{-z^2/2} = e^{-2.7220783^2/2} = .0246038$$

For m = 3 the z-score is

$$z = \sqrt{n}\, \frac{\bar{x} - m}{h} = \sqrt{32}\, \frac{2.406 - 3}{5} = -.6720343$$

And

$$e^{-z^2/2} = e^{-(-.6720343)^2/2} = .7978657$$

To get the likelihood, Minitab divides both .0246038 and .7978657 by the bigger of the two, .7978657, and multiplies by 1000000 (an arbitrary number—any other would do):

.0246038/.7978657x1000000 = 30837

and

.7978657/.7978657x1000000 = 1000000)

10.7 This is the Minitab input and output:

```
MTB > set 'm'
DATA> 395 400 405 410 415
DATA> end
MTB > set 'prior'
DATA> 1 1 1 1 1
DATA> end
MTB > exec 'm_disc'

INPUT POPULATION STANDARD DEVIATION:
DATA> 20

OBSERVED DATA IN WORKSHEET? (TYPE 'y' OR 'n'.)
  IF YES, INPUT NUMBER OF COLUMN.
  IF NO, INPUT OBSERVED SAMPLE MEAN AND SAMPLE SIZE:
n
DATA> 404.2 50

MEAN
  404.2

COUNT
  50

ROW   m  prior  M_x_PRIO    LIKE  PRODUCT     POST  M_x_POST

 1  395   0.2        79    5247     1049  0.003549     1.402
 2  400   0.2        80  345588    69118  0.233714    93.485
 3  405   0.2        81 1000000   200000  0.676278   273.892
 4  410   0.2        82  127137    25427  0.085980    35.252
 5  415   0.2        83     710      142  0.000480     0.199
 6                  405                              404.231
```

The posterior probabilities are shown in the column labeled "POST". This table from Minitab uses the scheme in which the largest likelihood is set to 1000000 by dividing by the largest likelihood throughout, and multiplying by 1000000. The table below gives the intermediate step in which the likelihoods are calculated directly from the following formulas:

$$z = \sqrt{n}\,\frac{\overline{x} - m}{h} \quad \text{and} \quad \text{likelihood} = e^{-z^2/2}$$

47

The so-called "Minitab likelihood" is calculated by dividing each of the "likelihood" entries by the greatest likelihood among the various models, in this case the one for m = 405, that is, .960789:

m	likelihood	Minitab likelihood
395	.005042	5247
400	.33204	345588
405	.960789	1000000
410	.122151	127137
415	.000682	710

The other columns in the Minitab table ("PRODUCT", etc.) are calculated in the usual way.

10.9 The mean of the 15 differences is 20.9. The input and output of Minitab are shown below.

(a) The posterior probabilities are shown in the column labeled "POST" in the Minitab output.

(b) A bar chart of the posterior probabilities is the following:

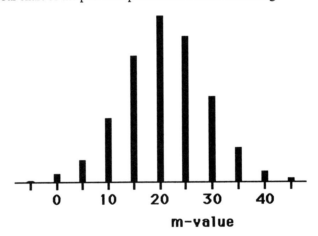

Minitab input and output (the entries into columns 1 are not shown here because I input them directly onto the Minitab Worksheet):

```
MTB > exec 'm_disc'

INPUT POPULATION STANDARD DEVIATION:
DATA> 30

OBSERVED DATA IN WORKSHEET? (TYPE 'y' OR 'n'.)
   IF YES, INPUT NUMBER OF COLUMN.
   IF NO, INPUT OBSERVED SAMPLE MEAN AND SAMPLE SIZE:
n
```

```
DATA> 20.93 15

MEAN
 20.93

COUNT
   15

ROW    m    prior  M_x_PRIO    LIKE    PRODUCT     POST    M_x_POST

  1  -100   0.025   -2.500        0        0.0   0.000000     0.0000
  2   -95   0.025   -2.375        0        0.0   0.000000     0.0000
  3   -90   0.025   -2.250        0        0.0   0.000000     0.0000
  4   -85   0.025   -2.125        0        0.0   0.000000     0.0000
  5   -80   0.025   -2.000        0        0.0   0.000000    -0.0000
  6   -75   0.025   -1.875        0        0.0   0.000000    -0.0000
  7   -70   0.025   -1.750        0        0.0   0.000000    -0.0000
  8   -65   0.025   -1.625        0        0.0   0.000000    -0.0000
  9   -60   0.025   -1.500        0        0.0   0.000000    -0.0000
 10   -55   0.025   -1.375        0        0.0   0.000000    -0.0000
 11   -50   0.025   -1.250        0        0.0   0.000000    -0.0000
 12   -45   0.025   -1.125        0        0.0   0.000000    -0.0000
 13   -40   0.025   -1.000        0        0.0   0.000000    -0.0000
 14   -35   0.025   -0.875        0        0.0   0.000000    -0.0000
 15   -30   0.025   -0.750        0        0.0   0.000000    -0.0000
 16   -25   0.025   -0.625        0        0.0   0.000000    -0.0000
 17   -20   0.025   -0.500        1        0.0   0.000000    -0.0000
 18   -15   0.025   -0.375       21        0.5   0.000005    -0.0001
 19   -10   0.025   -0.250      347        8.7   0.000089    -0.0009
 20    -5   0.025   -0.125     3713       92.8   0.000949    -0.0047
 21     0   0.025    0.000    26165      654.1   0.006690     0.0000
 22     5   0.025    0.125   121541     3038.5   0.031074     0.1554
 23    10   0.025    0.250   372196     9304.9   0.095158     0.9516
 24    15   0.025    0.375   751388    18784.7   0.192105     2.8816
 25    20   0.025    0.500  1000000    25000.0   0.255667     5.1133
 26    25   0.025    0.625   877364    21934.1   0.224313     5.6078
 27    30   0.025    0.750   507462    12686.6   0.129741     3.8922
 28    35   0.025    0.875   193496     4837.4   0.049470     1.7315
 29    40   0.025    1.000    48639     1216.0   0.012435     0.4974
 30    45   0.025    1.125     8060      201.5   0.002061     0.0927
 31    50   0.025    1.250      881       22.0   0.000225     0.0113
 32    55   0.025    1.375       63        1.6   0.000016     0.0009
 33    60   0.025    1.500        3        0.1   0.000001     0.0000
 34    65   0.025    1.625        0        0.0   0.000000     0.0000
 35    70   0.025    1.750        0        0.0   0.000000     0.0000
 36    75   0.025    1.875        0        0.0   0.000000     0.0000
 37    80   0.025    2.000        0        0.0   0.000000     0.0000
 38    85   0.025    2.125        0        0.0   0.000000     0.0000
 39    90   0.025    2.250        0        0.0   0.000000     0.0000
 40    95   0.025    2.375        0        0.0   0.000000     0.0000
 41                  -2.500                                  20.9300
```

10.11 The two m-values are now 1128 and 1136 and the mean of the single observation is 1120. The prior probabilities are now the posterior probabilities from the previous exercise. This is the input and output of Minitab:

```
MTB > set 'm'
DATA> 1128 1136
DATA> end
MTB > set 'prior'
DATA> .952574 .047426
DATA> end
MTB > exec 'm_disc'

INPUT POPULATION STANDARD DEVIATION:
DATA> 8

OBSERVED DATA IN WORKSHEET? (TYPE 'y' OR 'n'.)
  IF YES, INPUT NUMBER OF COLUMN.
  IF NO, INPUT OBSERVED SAMPLE MEAN AND SAMPLE SIZE:
n
DATA> 1120 1

MEAN
  1120

COUNT
  1

ROW   m     prior    M_x_PRIO   LIKE     PRODUCT     POST    M_x_POST

  1  1128  0.952574  1074.50   1000000   952574   0.989013   1115.61
  2  1136  0.047426    53.88    223130    10582   0.010987     12.48
  3                  1128.38                                 1128.09
```

The posterior probabilities are shown in the column labeled "POST". The probability of "Smith" has increased to about 99%.

CHAPTER 11
DENSITIES FOR MEANS

[In these answers I use the small-n correction factor regardless of the size of n. If you choose not to use this correction when n is large then your answer will be slightly different from mine. In some exercises I will use Minitab and in some I will not; when I use Minitab I will sometimes give a detailed solution as well.]

11.1 $\bar{x} = 21.14$, $s^2 = 16.26$, $s = 4.032$, n = 28, $s/\sqrt{n} = 4.03/\sqrt{28} = .762$

    (a) The posterior density is normal(21.14, .762).

$$z = \frac{20 - 21.14}{.762} = -1.49$$

    From the Standard Normal Table the probability that m is greater than 20 watts is about .932.

    (b) The predictive density is normal(21.14, $\sqrt{16.84}$) = normal(21.14, 4.10). The z-score is

$$z = \frac{20 - 21.14}{.410} = -.278$$

    The probability that the 29th observation will be greater that 20 watts is about .610.

11.3 $\bar{x} = 77.6$, $s^2 = 449.44$, $s = 21.2$, n = 5,
k = 1.8, h = sk = 38.16, $sk\sqrt{(n+1)/n} = 41.8$

The predictive density is normal(77.6, 41.8). The z-score is

$$z = \frac{150 - 77.6}{41.8} = 1.73$$

The probability that the 6th observation will be greater than 150 is about 4% (from the Standard Normal Table).

Using Minitab:

```
MTB > exec 'm_cont'

DO YOU WISH TO USE A FLAT PRIOR DENSITY FOR M? (TYPE 'y' OR 'n'.)
   IF NO, INPUT MEAN AND STANDARD DEVIATION FOR THE PRIOR DENSITY..
y
```

```
OBSERVED DATA IN WORKSHEET? (TYPE 'y' OR 'n'.)
  IF YES, INPUT NUMBER OF COLUMN.
  IF NO, INPUT OBSERVED SAMPLE MEAN, STANDARD DEVIATION, AND
SAMPLE SIZE.
y
DATA> 1

OBS_DATA
    73      75      49      76     115

MEAN
  77.6

STD
  21.2

COUNT
  5

THE POSTERIOR DENSITY FOR M IS NORMAL
WITH MEAN AND STANDARD DEVIATION:

MEAN STD
  77.6000   17.0657

THE PREDICTIVE DENSITY OF THE NEXT OBSERVATION
IS NORMAL WITH MEAN AND STANDARD DEVIATION:

MEAN STD
  77.6000   41.8022
```

11.5 $\bar{x} = 21.1$, $s^2 = 7.090$, $s = 2.663$, $k = 1.2$, $h = sk = 3.195$, $sk\sqrt{(n+1)/n} = 3.351$

The predictive density is normal(21.1, 3.351).

(a) The z-score is

$$z = \frac{29 - 21.1}{3.351} = 2.36$$

From the Standard Normal Table, the probability that the next observation will be greater than 29 is about .009.

(b) The z-score is

$$z = \frac{14 - 21.1}{3.351} = -2.12$$

So the probability that the next observation will be less than 14 is about .017.

Using Minitab:

```
MTB > exec 'm_cont'

DO YOU WISH TO USE A FLAT PRIOR DENSITY FOR M? (TYPE 'y' OR 'n'.)
   IF NO, INPUT MEAN AND STANDARD DEVIATION FOR THE PRIOR DENSITY..
y

OBSERVED DATA IN WORKSHEET? (TYPE 'y' OR 'n'.)
   IF YES, INPUT NUMBER OF COLUMN.
   IF NO, INPUT OBSERVED SAMPLE MEAN, STANDARD DEVIATION, AND
SAMPLE SIZE.
n
DATA> 21.1 2.663 10
MEAN
  21.1

STD
  2.663

COUNT
  10

THE POSTERIOR DENSITY FOR M IS NORMAL
WITH MEAN AND STANDARD DEVIATION:
MEAN STD
  21.1000     1.0105

THE PREDICTIVE DENSITY OF THE NEXT OBSERVATION
IS NORMAL WITH MEAN AND STANDARD DEVIATION:
MEAN STD
  21.1000     3.3516
```

11.7 $\bar{x} = 11.00$, $s = 4.913$, $n = 30$, $s/\sqrt{n} = .8969$, $k = 1.022$, $sk/\sqrt{n} = .9169$

The posterior density of the mean difference is normal(4.233, .8921).

(**a**) The z-score is

$$z = \frac{0 - 11.00}{.9169} = -12.00$$

The posterior probability that the mean difference is positive is the probability that $z > -12$, which is 1.0000.

(b) The z-score for difference x is

$$z = \frac{x - 11.00}{.9169}$$

The following table gives the z-scores for the corresponding x's. The various PdALx are calculated from the Standard Normal Table.

x	2	4	6	8
z-score	-9.82	-7.63	-5.45	3.27
PdALx	1.000	1.000	1.000	.999

Using Minitab:

```
MTB > exec 'm_cont'

DO YOU WISH TO USE A FLAT PRIOR DENSITY FOR M? (TYPE 'y' OR 'n'.)
  IF NO, INPUT MEAN AND STANDARD DEVIATION FOR THE PRIOR DENSITY..
y

OBSERVED DATA IN WORKSHEET? (TYPE 'y' OR 'n'.)
  IF YES, INPUT NUMBER OF COLUMN.
  IF NO, INPUT OBSERVED SAMPLE MEAN, STANDARD DEVIATION, AND
SAMPLE SIZE.
n
DATA> 11 4.913 30
MEAN
   11

STD
   4.913

COUNT
   30

THE POSTERIOR DENSITY FOR M IS NORMAL
WITH MEAN AND STANDARD DEVIATION:
MEAN STD
  11.0000    0.9169

THE PREDICTIVE DENSITY OF THE NEXT OBSERVATION
IS NORMAL WITH MEAN AND STANDARD DEVIATION:
MEAN STD
  11.0000    5.1052
```

11.9 $m_0 = 6$, $z = \frac{12 - 6}{h_0} = 1.04$, so $h_0 = 6/1.04 = 5.77$

11.11 $m_0 = 9$, $z = \frac{20 - 9}{h_0} = .674$, so $h_0 = 11/.674 = 16.32$

The z-score for m = 0 is

$$z = \frac{0 - m_0}{h_0} = -.55$$

The probability of $m < 0$ is .29. Since it is impossible for a future earthquake to take place in the past, the actual probability of $m < 0$ is 0. Therefore a normal density cannot apply exactly.

11.13 $\bar{x} = 14.38$, $n = 21$, $s = 7.755$, $m_0 = 12$, $h_0 = 6$, $k = 1.045$, $h = sk = 8.107$

(a) Prior precision: $c_0 = 1/h_0^2 = 1/36 = .0278$

Sample precision: $c = n/h^2 = .31954$

Posterior precision: $c_1 = c_0 + c = 1/36 + .3195 = .3473$

Posterior mean:

$$m_1 = \frac{c_0}{c_1}m_0 + \frac{c}{c_1}\bar{x} = \frac{.0278}{.3473}m_0 + \frac{.3195}{.3473}\bar{x} = .0800(12) = .9200(14.38) = 14.19$$

Posterior standard deviation: $h_1 = 1/\sqrt{c_1} = h/\sqrt{n} = 8.107/\sqrt{21} = 1.769$

Posterior density of m is normal(14.19, 1.769)

(b) Standard deviation of predictive density is $\sqrt{h^2 + 1/c_1} = \sqrt{8.107 + 1/.3473} = 8.282$.

Predictive density is normal(14.19, 8.282)

The posterior probability that a child has not uttered a word at age 24 months is the probability (from the Standard Normal Table) to the right of the following z-score:

$$z = \frac{24 - 14.19}{8.282} = 1.184$$

The probability to the right of $z = 1.18$ is about 11.8%.

11.15 $\bar{x} = 2.41$, $s = 4.44$, $n = 32$, $k = 1 + 20/32^2 = 1.020$, $h = sk = 4.53$, $m_0 = 0$, $h_0 = 10$, $c_0 = 1/h_0^2 = .01$, $c = n/h^2 = 1.56$, $c_1 = c_0 + c = 1.57$, $m_1 = \frac{c_0}{c_1}m_0 + \frac{c}{c_1}\bar{x} = 2.39$

The posterior density of m is normal(2.39, .800). The predictive density of the next observation is normal($m_1, \sqrt{h^2 + 1/c_1}$) = normal(2.39, 4.60).

For the predictive probability of the next patient will have a positive difference, calculate the z-score at 0:

$$z = \frac{0 - 2.39}{4.60} = -.52$$

According to the Standard Normal Table, the probability to the right in this is about 70%.

Using Minitab:

```
MTB > exec 'm_cont'

DO YOU WISH TO USE A FLAT PRIOR DENSITY FOR M? (TYPE 'y' OR 'n'.)
   IF NO, INPUT MEAN AND STANDARD DEVIATION FOR THE PRIOR DENSITY..
n
DATA> 0 10
PR_MEAN
   0

PR_STD
   10

OBSERVED DATA IN WORKSHEET? (TYPE 'y' OR 'n'.)
   IF YES, INPUT NUMBER OF COLUMN.
   IF NO, INPUT OBSERVED SAMPLE MEAN, STANDARD DEVIATION, AND
SAMPLE SIZE.
n
DATA> 2.41 4.44 20
MEAN
   2.41

STD
   4.44

COUNT
   20

THE POSTERIOR DENSITY FOR M IS NORMAL
WITH MEAN AND STANDARD DEVIATION:
MEAN STD
   2.38409   1.03684

THE PREDICTIVE DENSITY OF THE NEXT OBSERVATION
IS NORMAL WITH MEAN AND STANDARD DEVIATION:
MEAN STD
   2.38409   4.77591
```

For the predictive probability of the next patient will have a positive difference, calculate

$$z = \frac{0 - 2.384}{4.776} = -.50$$

The probability to the right of $z = -.50$ is (from the Standard Normal Table) about 69%.

11.17 The mean of the sample of 208 is 53.60, and the standard deviation is s = 15.18.

Using Minitab:

```
MTB > exec 'm_cont'

DO YOU WISH TO USE A FLAT PRIOR DENSITY FOR M? (TYPE 'y' OR 'n'.)
  IF NO, INPUT MEAN AND STANDARD DEVIATION FOR THE PRIOR DENSITY..
n
DATA> 60 20
PR_MEAN
   60

PR_STD
   20

OBSERVED DATA IN WORKSHEET? (TYPE 'y' OR 'n'.)
  IF YES, INPUT NUMBER OF COLUMN.
  IF NO, INPUT OBSERVED SAMPLE MEAN, STANDARD DEVIATION, AND
SAMPLE SIZE.
n
DATA> 53.60 15.18 208
MEAN
   53.6

STD
   15.18

COUNT
   208

THE POSTERIOR DENSITY FOR M IS NORMAL
WITH MEAN AND STANDARD DEVIATION:
MEAN STD
  53.6177    1.0516

THE PREDICTIVE DENSITY OF THE NEXT OBSERVATION
IS NORMAL WITH MEAN AND STANDARD DEVIATION:
MEAN STD
  53.6177    15.2234
```

(a) The probability the next observation is greater than 70 is the probability to the right of $z = (70 - 53.62)/15.22 = 1.08$, which according to the Standard Normal Table is about 14%.

(b) The 13th largest score among the 208 in the sample is 82. The probability the next observation is at least as large as 82 is the probability to the right of $z = (82 - 53.62)/15.22 = 1.86$, which according to the Standard Normal Table is about 3.1%.

(c) The probability the next observation is at least as large as 82 given that it is greater than 70 is the ratio of the answers to (b) and (a):

$$P(\geq 82 \mid > 70) = \frac{P(\geq 82 \cap > 70)}{P(> 70)} = \frac{P(\geq 82)}{P(> 70)} = \frac{.031}{.14} = 22\%$$

11.19 (a) From Example 11.2 the posterior density of m is normal(18.25, .284). A 99% posterior probability interval is $18.25 \pm 2.58 \times .284$, or from about 17.52 to 18.98.

(b) From Exercise 11.1 the posterior density of m is normal(21.14, .762). A 99% posterior probability interval is $21.14 \pm 2.58 \times .762$, or from about 19.17 to 23.11.

11.21 $\bar{x} = 21.6$ and $s = 4.05$ from Exercise 2.22

99% probability interval for m is $21.6 \pm 2.58 \times (1+20/169) \times 4.05/\sqrt{13} = 21.6 \pm 3.24$, or from about 18.4 to 24.8.

Using Minitab:

```
MTB > exec 'm_cont'

DO YOU WISH TO USE A FLAT PRIOR DENSITY FOR M? (TYPE 'y' OR 'n'.)
  IF NO, INPUT MEAN AND STANDARD DEVIATION FOR THE PRIOR DENSITY..
y
OBSERVED DATA IN WORKSHEET? (TYPE 'y' OR 'n'.)
  IF YES, INPUT NUMBER OF COLUMN.
  IF NO, INPUT OBSERVED SAMPLE MEAN, STANDARD DEVIATION, AND
SAMPLE SIZE.
n
DATA> 21.6 4.05 13

MEAN
  21.6

STD
  4.05

COUNT
   13

THE POSTERIOR DENSITY FOR M IS NORMAL
WITH MEAN AND STANDARD DEVIATION:

MEAN STD
 21.6000    1.2562

THE PREDICTIVE DENSITY OF THE NEXT OBSERVATION
IS NORMAL WITH MEAN AND STANDARD DEVIATION:

MEAN STD
 21.6000    4.7003
```

11.23 $\bar{x} = .422$, $s = .24$, $n = 18$, $k = 1+20/18^2 = 1.062$, $sk/\sqrt{n} = .0601$

The posterior density of m is normal(.422, .0601). A 95% posterior probability interval for m is $.422 \pm 1.96 \times .0601$, or from about .304 to .540.

11.25 $\bar{x} = 7.337$, $s = 1.292$, $n = 43$

(a) $m_0 = 7.5$, $h_0 = 2.0$. The posterior density for m is normal(7.339, .1982). A 90% probability interval for m is $7.339 \pm 1.65 \times .1982 = 7.339 \pm .327$, or from about 7.012 to 7.666.

(b) The posterior density for m is normal(7.337, .1992). A 90% probability interval for m is $7.337 \pm 1.65 \times .1992 = 7.337 \pm .329$, or from about 7.008 to 7.666. (This is the same as that in part (a) to two decimal points, which indicates that the prior precision in (a) is quite small in comparison with the sample precision.)

11.27 (a) The posterior density for m_{DDT} is normal(2.7, 1.22). A 95% probability interval for the population mean difference m_{DDT} is $2.7 \pm 1.96 \times 1.22 = 2.7 \pm 2.4$, or from about .3 to 5.1. Since this interval does not contain $m_{DDT} = 0$, the null hypothesis of no difference is not supported—the evidence supports the conclusion that the blood of breast cancer patients contains elevated levels of DDT, on the average.

(b) The posterior density for m_{PCB} is normal(1.0, .52). A 95% probability interval for the population mean difference m_{PCB} is $1.0 \pm 1.96 \times .52 = 1.0 \pm 1.01$, or from about -.01 to 2.01. Since this interval contains $m_{PCB} = 0$, the null hypothesis of no difference is supported (but barely)—the evidence does provide very strong support to the conclusion that the blood of breast cancer patients contains elevated levels of PCBs, on the average.

CHAPTER 12
COMPARING TWO OR MORE MEANS

[In these answers I use the small-n correction factor regardless of the size of n. If you choose not to use this correction when n is large then your answer will be slightly different from mine. For some solutions I will use Minitab and for some I will not; with Minitab output I will sometimes give a detailed solution as well. To save space, I will not reproduce all of Minitab's output.]

12.1 The density of d is normal(2.45, .98). The z-score is

$$z = \frac{0 - 2.45}{.98} = -2.50$$

The posterior probability of d > 0 is the probability to the right of this value of z, which according to the Standard Normal Table is about .994.

12.3 For the Intervention group: $\bar{x}_I = 6.07$, $s_I = 4.11$, $n_I = 13$, $k_I = 1+20/13^2 = 1.12$. The posterior density of m_I is normal(6.07, 1.27).

For the Control group: $\bar{x}_C = 8.60$, $s_C = 2.41$, $n_C = 10$, $k_C = 1+20/10^2 = 1.2$. The posterior density of m_C is normal(8.60, .91).

The posterior density of $d = m_I - m_C$ is normal(6.07 - 8.60, $\sqrt{1.27^2+.91^2}$) = normal(-2.53, 1.57).

(a) The z-score is

$$z = \frac{0 - (-2.53)}{1.57} = 1.61$$

The probability that the Intervention is effective is 1 - PdAL0 = .946, from the Standard Normal Table.

(b) A 90% posterior probability interval for d is -2.53 ± 1.65x1.57 = -2.53 ± 2.59, or from about -5.12 to .06.

(c) A 95% posterior probability interval for d is -2.53 ± 1.96x1.57 = -2.53 ± 3.08, or from about -5.61 to .55. Since this interval contains 0, the null hypothesis that d = 0 is supported.

12.5 For the Formoterol group: $\bar{x}_F = 27.2$, $s_F = 6.063$, $n_F = 30$, $k_I = 1+20/30^2 = 1.022$. The posterior density of m_F is normal(27.2, 1.132).

For the Salbutamol group: $\bar{x}_S = 22.97$, $s_S = 5.57$, $n_S = 30$, $k_S = 1+20/30^2 = 1.022$. The posterior density of m_F is normal(22.97, 1.0395).

The posterior density of $d = m_F - m_S$ is normal(27.2-22.97, $\sqrt{1.13^2+1.04^2}$) = normal(4.23, 1.54).

In the following table, $z = (x - 4.23)/1.54$, and PdALx is from the Standard Normal Table:

x	0	2	4	6	8
z-score	-2.75	-1.45	-.15	1.15	2.45
PdALx	.997	.926	.560	.125	.007

As expected, these are quite similar to the entries in the table in Exercise 11.6.

12.7 From Exercise 11.18, the posterior density for m_1 is normal(122.6, 4.76).

Regarding sample 2: $\bar{x}_2 = 97$, $s_2 = 12$, $n_2 = 2$, $k_2 = 1+20/2^2 = 6$. The posterior density of m_2 is normal(99.6, 18.62).

(a) The posterior density of $d = m_1 - m_2$ is normal(122.6 - 99.6, $\sqrt{4.76^2+18.6^2}$) = normal(23.0, 19.2).

(b) The z-score for finding PdALx has formula $z = (x - 23.0)/19.2$.
For $x = 0$, $z = (0 - 23.0)/19.2 = -1.20$; from the Standard Normal Table, PdAL0 = .885.
For $x = 10$, $z = (10 - 23.0)/19.2 = -.68$; from the Standard Normal Table, PdAL10 = .752.

(c) The 95% posterior probability interval for d is $23.0 \pm 1.96 \times 19.2 = 23.0 \pm 37.7$, or from about -14.7 to 60.7.

(d) The interval in (c) contains 0 and so the null hypothesis of no difference between the mean weights of the two specimens ($d = 0$) is supported. (Since there are only two weights from specimen 2, the only way one could safely conclude otherwise would be if there was substantial prior information about m_2, which is not the case in this exercise.)

12.9 For cannula A: $\bar{x}_A = .7306$, $s_A = .3881$, $n_A = 16$, $k_A = 1+20/16^2 = 1.08$. The posterior density of m_A is normal(.731, .105).

For cannula B: $\bar{x}_B = .3656$, $s_B = .1726$, $n_B = 16$, $k_B = 1+20/16^2 = 1.08$. The posterior density of m_B is normal(.366, .0465).

The posterior density of $d = m_A - m_B$ is normal(.731-.366, $\sqrt{.105^2+.0465^2}$) = normal(.365, .115). A 95% posterior probability interval is $.365 \pm 1.96 \times .115 = .365 \pm .225$, or from about .140 to .590. Since this interval does not contain the null value, d = 0, the evidence supports the conclusion that the population mean of A is greater than that of B.

12.11 Using Minitab for Machine A: $m_0 = 5.65$, $h_0 = .1$, $\bar{x}_A = 5.63$, $s_A = .025$, $n_A = 20$:

```
MTB > exec 'm_cont'

INPUT MEAN AND STANDARD DEVIATION FOR THE PRIOR DENSITY..
DATA> 5.65 .1

INPUT OBSERVED SAMPLE MEAN, STANDARD DEVIATION, AND SAMPLE SIZE.
DATA> 5.6345 .025 20

THE POSTERIOR DENSITY FOR M IS NORMAL
WITH MEAN AND STANDARD DEVIATION:
  5.63455    0.00586
```

Using Minitab for Machine B: $m_0 = 5.65$, $h_0 = .1$, $\bar{x}_B = 5.63$, $s_B = .0264$, $n_B = 25$:

```
MTB > exec 'm_cont'

INPUT MEAN AND STANDARD DEVIATION FOR THE PRIOR DENSITY..
DATA> 5.65 .1

INPUT OBSERVED SAMPLE MEAN, STANDARD DEVIATION, AND SAMPLE SIZE.
DATA> 5.6340 .02638 25

THE POSTERIOR DENSITY FOR M IS NORMAL
WITH MEAN AND STANDARD DEVIATION:
  5.63405    0.00544
```

For the difference in means:

```
MTB > exec 'mm_cont'

INPUT MEAN AND STANDARD DEVIATION FOR NORMAL DISTRIBUTION FOR MEAN
M1:
DATA> 5.63455 .00586

INPUT MEAN AND STANDARD DEVIATION FOR NORMAL DISTRIBUTION FOR MEAN
M2:
DATA> 5.63405 0.00544

THE POSTERIOR DENSITY FOR M1-M2 IS NORMAL
WITH MEAN AND STANDARD DEVIATION:
  0.0005002    0.0079958
```

A 95% posterior probability interval for d is $.0005 \pm 1.96 \times .008 = .0005 \pm .0157$, or from about -.015 to .016.

12.13 First find the density of m_T, the mean for trained octopuses, using Minitab:

```
MTB > exec 'm_cont'

DO YOU WISH TO USE A FLAT PRIOR DENSITY FOR M? (TYPE 'y' OR 'n'.)
y
```

```
INPUT OBSERVED SAMPLE MEAN, STANDARD DEVIATION, AND SAMPLE SIZE.
DATA> .862 .196 16

THE POSTERIOR DENSITY FOR M IS NORMAL
WITH MEAN AND STANDARD DEVIATION:
   0.862000   0.052828
```

Now for the density of m_C, the mean of the Controls:

```
MTB > exec 'm_cont'

DO YOU WISH TO USE A FLAT PRIOR DENSITY FOR M? (TYPE 'y' OR 'n'.)
y

INPUT OBSERVED SAMPLE MEAN, STANDARD DEVIATION, AND SAMPLE SIZE.
DATA> .422 .240 18

THE POSTERIOR DENSITY FOR M IS NORMAL
WITH MEAN AND STANDARD DEVIATION:
   0.422000   0.060060
```

Considering the difference:

```
MTB > exec 'mm_cont'

INPUT MEAN AND STANDARD DEVIATION FOR NORMAL DISTRIBUTION FOR MEAN
M1:
DATA> 0.862000   0.052828

INPUT MEAN AND STANDARD DEVIATION FOR NORMAL DISTRIBUTION FOR MEAN
M2:
DATA> 0.422000   0.060060

THE POSTERIOR DENSITY FOR M1-M2 IS NORMAL
WITH MEAN AND STANDARD DEVIATION:
   0.44   0.0799875
```

(**a**) To find the probability that $d = m_T - m_C > 0$ use the z-score at $d = 0$:

$$z = \frac{0 - .44}{.0800} = -5.50$$

The probability to the right of this z is 1.000.

(**b**) A 68% probability interval for d is $.44 \pm 1.00 \times .080 = .44 \pm .08$, or from about .36 to .52.

(**c**) A 95% probability interval for d is $.44 \pm 1.96 \times .080 = .44 \pm .16$, or from about .28 to .60. This interval does not contain 0, indicating that the null hypothesis $d = 0$ is not supported. Nor is 0 contained in a 99% interval (.23 to .65). The evidence is reasonably strong that octopuses can be trained.

12.15 Using Minitab for m_M:

```
MTB > exec 'm_cont'

DO YOU WISH TO USE A FLAT PRIOR DENSITY FOR M? (TYPE 'y' OR 'n'.)
y

OBSERVED DATA IN WORKSHEET? (TYPE 'y' OR 'n'.)
  IF YES, INPUT NUMBER OF COLUMN.
y
DATA> 1

OBS_DATA
  35.0   36.8   40.2   46.6   50.4   64.2   83.0   87.6   89.2

MEAN
  59.2222

STD
  21.0359

COUNT
   9

THE POSTERIOR DENSITY FOR M IS NORMAL
WITH MEAN AND STANDARD DEVIATION:
  59.2222    8.7433
```

Using Minitab for m_N:

```
MTB > exec 'm_cont'

DO YOU WISH TO USE A FLAT PRIOR DENSITY FOR M? (TYPE 'y' OR 'n'.)
y

OBSERVED DATA IN WORKSHEET? (TYPE 'y' OR 'n'.)
  IF YES, INPUT NUMBER OF COLUMN.
y
DATA> 1

OBS_DATA
  28.2   28.6   33.0   34.8   45.4   50.8   52.6   66.4   67.8

MEAN
  45.2889

STD
  14.4062

COUNT
   9

THE POSTERIOR DENSITY FOR M IS NORMAL
WITH MEAN AND STANDARD DEVIATION:
  45.2889    5.9878
```

Now the difference:

```
MTB > exec 'mm_cont'

INPUT MEAN AND STANDARD DEVIATION FOR NORMAL DISTRIBUTION FOR MEAN
M1:
DATA> 59.2222    8.7433

INPUT MEAN AND STANDARD DEVIATION FOR NORMAL DISTRIBUTION FOR MEAN
M2:
DATA> 45.2889    5.9878

THE POSTERIOR DENSITY FOR M1-M2 IS NORMAL
WITH MEAN AND STANDARD DEVIATION:
   13.9333   10.5971
```

To test the null hypothesis d = 0 find the 95% probability interval for d: $13.9 \pm 1.96 \times 10.6 = 13.9 \pm 20.8$, or from about -6.8 to 34.7. Since 0 is contained in this interval the null hypothesis is supported.

12.17 Using Minitab for m_G:

```
MTB > exec 'm_cont'

DO YOU WISH TO USE A FLAT PRIOR DENSITY FOR M? (TYPE 'y' OR 'n'.)
y

OBSERVED DATA IN WORKSHEET? (TYPE 'y' OR 'n'.)
  IF NO, INPUT OBSERVED SAMPLE MEAN, STANDARD DEVIATION, AND
SAMPLE SIZE.
n
DATA> 14.2033 3.44563 30

THE POSTERIOR DENSITY FOR M IS NORMAL
WITH MEAN AND STANDARD DEVIATION:
   14.2033    0.6431
```

Using Minitab for m_S:

```
MTB > exec 'm_cont'

DO YOU WISH TO USE A FLAT PRIOR DENSITY FOR M? (TYPE 'y' OR 'n'.)
y

OBSERVED DATA IN WORKSHEET? (TYPE 'y' OR 'n'.)
  IF NO, INPUT OBSERVED SAMPLE MEAN, STANDARD DEVIATION, AND
SAMPLE SIZE.
n
DATA> 10.6133 2.59586 30

THE POSTERIOR DENSITY FOR M IS NORMAL
WITH MEAN AND STANDARD DEVIATION:
   10.6133    0.4845
```

Now for the difference d = m_G - m_S:

```
MTB > exec 'mm_cont'

INPUT MEAN AND STANDARD DEVIATION FOR NORMAL DISTRIBUTION FOR MEAN
M1:
DATA> 14.2033    0.6431

INPUT MEAN AND STANDARD DEVIATION FOR NORMAL DISTRIBUTION FOR MEAN
M2:
DATA> 10.6133    0.4845

THE POSTERIOR DENSITY FOR M1-M2 IS NORMAL
WITH MEAN AND STANDARD DEVIATION:
   3.59    0.805182
```

(a) The 95% posterior probability interval for d is 3.59 ± 1.96x.805 = 3.59 ± 1.58, or from about 2.01 to 5.17. The high probability that d is greater than 0 supports the conclusion that there is a genetic component to sexual orientation.

(b) The interval in (a) does not contain d = 0 and so the evidence does not support the null hypothesis.

12.19 Using Minitab for m_A:

```
MTB > exec 'm_cont'

DO YOU WISH TO USE A FLAT PRIOR DENSITY FOR M? (TYPE 'y' OR 'n'.)
y

OBSERVED DATA IN WORKSHEET? (TYPE 'y' OR 'n'.)
  IF NO, INPUT OBSERVED SAMPLE MEAN, STANDARD DEVIATION, AND
SAMPLE SIZE.
n
DATA> 2.31013 .000132 8

THE POSTERIOR DENSITY FOR M IS NORMAL
WITH MEAN AND STANDARD DEVIATION:
   2.31013    0.00006
```

Using Minitab for m_C:

```
MTB > 12.19

MTB > exec 'm_cont'

DO YOU WISH TO USE A FLAT PRIOR DENSITY FOR M? (TYPE 'y' OR 'n'.)
y

OBSERVED DATA IN WORKSHEET? (TYPE 'y' OR 'n'.)
  IF NO, INPUT OBSERVED SAMPLE MEAN, STANDARD DEVIATION, AND
SAMPLE SIZE.
n
DATA> 2.29947 .00129 8
```

```
THE POSTERIOR DENSITY FOR M IS NORMAL
WITH MEAN AND STANDARD DEVIATION:
  2.29947   0.00060
```

Using Minitab to find the posterior density of d:

```
MTB > exec 'mm_cont'

INPUT MEAN AND STANDARD DEVIATION FOR NORMAL DISTRIBUTION FOR MEAN
M2:
DATA> 2.31013   0.00006

INPUT MEAN AND STANDARD DEVIATION FOR NORMAL DISTRIBUTION FOR MEAN
M1:
DATA> 2.29947   0.00060

THE POSTERIOR DENSITY FOR M1-M2 IS NORMAL
WITH MEAN AND STANDARD DEVIATION:
  0.0106599   0.0006030
```

(a) The 95% probability interval for d is .01066 ± 1.96x.00060 = .01066 ± .00118, or from about .00948 to .01184.

(b) The interval in (a) does not include 0 and so the null hypothesis of no difference in the two means is not supported. Actually, the interval is very far from 0 relative to its size—as shown below—and so the evidence is quite strong that these two methods are measuring something different.

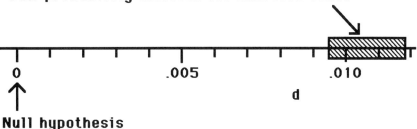

12.21 (a) Using Minitab for m_I:

```
MTB > exec 'm_cont'

DO YOU WISH TO USE A FLAT PRIOR DENSITY FOR M? (TYPE 'y' OR
'n'.)
  IF NO, INPUT MEAN AND STANDARD DEVIATION FOR THE PRIOR
DENSITY..
n
DATA> 40 10

OBSERVED DATA IN WORKSHEET? (TYPE 'y' OR 'n'.)
  IF NO, INPUT OBSERVED SAMPLE MEAN, STANDARD DEVIATION, AND
SAMPLE SIZE.
n
DATA> 32.9 8.2 156
```

```
THE POSTERIOR DENSITY FOR M IS NORMAL
WITH MEAN AND STANDARD DEVIATION:
  32.9305    0.6557
```

Using Minitab for m_C:

```
MTB > exec 'm_cont'

DO YOU WISH TO USE A FLAT PRIOR DENSITY FOR M? (TYPE 'y' OR
'n'.)
  IF NO, INPUT MEAN AND STANDARD DEVIATION FOR THE PRIOR
DENSITY..
n
DATA> 40 10

OBSERVED DATA IN WORKSHEET? (TYPE 'y' OR 'n'.)
  IF NO, INPUT OBSERVED SAMPLE MEAN, STANDARD DEVIATION, AND
SAMPLE SIZE.
n
DATA> 36.9 7.9 148

THE POSTERIOR DENSITY FOR M IS NORMAL
WITH MEAN AND STANDARD DEVIATION:
  36.9130    0.6486
```

Using Minitab for the difference:

```
MTB > exec 'mm_cont'

INPUT MEAN AND STANDARD DEVIATION FOR NORMAL DISTRIBUTION FOR
MEAN M1:
DATA> 32.9305    0.6557

INPUT MEAN AND STANDARD DEVIATION FOR NORMAL DISTRIBUTION FOR
MEAN M2:
DATA> 36.9130    0.6486

THE POSTERIOR DENSITY FOR M1-M2 IS NORMAL
WITH MEAN AND STANDARD DEVIATION:
  -3.9825    0.922293
```

The 95% posterior probability interval for $d = m_I - m_C$ is $-3.98 \pm 1.96 \times .92 =$ -3.98 ± 1.81, or from about -5.79 to -2.17. Since this interval does not contain $d = 0$, the evidence does not support the null hypothesis.

(**b**) Using Minitab for m_I:

```
MTB > exec 'm_cont'

DO YOU WISH TO USE A FLAT PRIOR DENSITY FOR M? (TYPE 'y' OR
'n'.)
  IF NO, INPUT MEAN AND STANDARD DEVIATION FOR THE PRIOR
DENSITY..
n
DATA> 60 10
```

```
OBSERVED DATA IN WORKSHEET? (TYPE 'y' OR 'n'.)
   IF NO, INPUT OBSERVED SAMPLE MEAN, STANDARD DEVIATION, AND
SAMPLE SIZE.
n
DATA> 54 20 156

THE POSTERIOR DENSITY FOR M IS NORMAL
WITH MEAN AND STANDARD DEVIATION:
   54.1502   1.5824
```

Using Minitab for m_C:

```
MTB > exec 'm_cont'

DO YOU WISH TO USE A FLAT PRIOR DENSITY FOR M? (TYPE 'y' OR
'n'.)
   IF NO, INPUT MEAN AND STANDARD DEVIATION FOR THE PRIOR
DENSITY..
n
DATA> 60 10

OBSERVED DATA IN WORKSHEET? (TYPE 'y' OR 'n'.)
   IF NO, INPUT OBSERVED SAMPLE MEAN, STANDARD DEVIATION, AND
SAMPLE SIZE.
n
DATA> 58 22 148

THE POSTERIOR DENSITY FOR M IS NORMAL
WITH MEAN AND STANDARD DEVIATION:
   58.0634   1.7811
```

Using Minitab for the difference $d = m_I - m_C$:

```
MTB > exec 'mm_cont'

INPUT MEAN AND STANDARD DEVIATION FOR NORMAL DISTRIBUTION FOR
MEAN M1:
DATA> 54.1502   1.5824

INPUT MEAN AND STANDARD DEVIATION FOR NORMAL DISTRIBUTION FOR
MEAN M2:
DATA> 58.0634   1.7811

THE POSTERIOR DENSITY FOR M1-M2 IS NORMAL
WITH MEAN AND STANDARD DEVIATION:
   -3.9132   2.3825
```

The 95% posterior probability interval for $d = m_I - m_C$ is $-3.91 \pm 1.96 \times 2.38 = -3.91 \pm 4.67$, or from about -8.58 to $+.76$. Since this interval contains $d = 0$, the evidence supports the null hypothesis.

12.23 For mother smokes or quit: $\bar{x}_T = 6.82$, $s_T = .98$, $n_T = 29$, $m_{T0} = 7.1$, $h_{T0} = 2$,

$$k_T = 1 + 20/n_T^2 = 1.024, \quad h_T = s_T k_T = 1.00, \quad c_{T0} = 1/h_{T0}^2 = .25,$$

$$c_T = n_T/h_T = 28.8, \quad c_{T1} = c_{T0} + c_T = 29.1, \quad m_{T1} = \frac{c_{T0}}{c_{T1}}m_{T0} + \frac{c_T}{c_{T1}}\bar{x}_T = 6.82$$

For mother never smoked: $\bar{x}_C = 7.34$, $s_C = 1.29$, $n_C = 43$, $m_{C0} = 7.5$, $h_{C0} = 2$,

$$k_C = 1 + 20/n_C^2 = 1.011, \quad h_C = s_C k_C = 1.30, \quad c_{C0} = 1/h_{C0}^2 = .25,$$

$$c_C = n_C/h_C = 25.3, \quad c_{C1} = c_{C0} + c_C = 25.5, \quad m_{C1} = \frac{c_{C0}}{c_{C1}}m_{C0} + \frac{c_C}{c_{C1}}\bar{x}_C = 7.34$$

Posterior density of d: normal$(m_{T1} - m_{C1}, \sqrt{1/c_{T1} + 1/c_{C1}})$ = normal$(-.52, .271)$

(**a**) The z-score for $x = 0$ is

$$z = \frac{0 - (-.52)}{.271} = 1.91$$

So PdAL0 is about .028.

(**b**) The 90% probability interval for d is $-.52 \pm 1.65 \times .271 = -.52 \pm .45$, or from about $-.97$ to $-.07$.

(**c**) The 95% probability interval for d is $-.52 \pm 1.96 \times .271 = -.52 \pm .53$, or from about -1.05 to $.01$. Since this interval contains 0 (but barely!), the evidence supports the null hypothesis of no difference in birth weights when the mothers smoke and not.

Alternatively, using Minitab to find the posterior density of d:

```
MTB > exec 'm_cont'

DO YOU WISH TO USE A FLAT PRIOR DENSITY FOR M? (TYPE 'y' OR 'n'.)
   IF NO, INPUT MEAN AND STANDARD DEVIATION FOR THE PRIOR DENSITY..
n
DATA> 7.1 2

OBSERVED DATA IN WORKSHEET? (TYPE 'y' OR 'n'.)
   IF NO, INPUT OBSERVED SAMPLE MEAN, STANDARD DEVIATION, AND
SAMPLE SIZE.
n
DATA> 6.82 .98 29

THE POSTERIOR DENSITY FOR M IS NORMAL
WITH MEAN AND STANDARD DEVIATION:
   6.82241    0.18551

MTB > exec 'm_cont'

DO YOU WISH TO USE A FLAT PRIOR DENSITY FOR M? (TYPE 'y' OR 'n'.)
   IF NO, INPUT MEAN AND STANDARD DEVIATION FOR THE PRIOR DENSITY..
n
DATA> 7.5 2
```

```
OBSERVED DATA IN WORKSHEET? (TYPE 'y' OR 'n'.)
   IF NO, INPUT OBSERVED SAMPLE MEAN, STANDARD DEVIATION, AND
SAMPLE SIZE.
n
DATA> 7.34 1.29 43

THE POSTERIOR DENSITY FOR M IS NORMAL
WITH MEAN AND STANDARD DEVIATION:
   7.34157   0.19788

MTB > exec 'mm_cont'

INPUT MEAN AND STANDARD DEVIATION FOR NORMAL DISTRIBUTION FOR MEAN
M1:
DATA> 6.82241   0.18551

INPUT MEAN AND STANDARD DEVIATION FOR NORMAL DISTRIBUTION FOR MEAN
M2:
DATA> 7.34157   0.19788

THE POSTERIOR DENSITY FOR M1-M2 IS NORMAL
WITH MEAN AND STANDARD DEVIATION:
   -0.51916   0.271239
```

12.25 The relevant calculations are shown in the following table:

	k	h	m_1	h_1
Exercise Yes:	1.0113	44.499	28.4	6.8663
Exercise No:	1.0119	50.595	3.6	7.9016

Or, using Minitab:

```
MTB > exec 'm_cont'

DO YOU WISH TO USE A FLAT PRIOR DENSITY FOR M? (TYPE 'y' OR 'n'.)
y

OBSERVED DATA IN WORKSHEET? (TYPE 'y' OR 'n'.)
   IF NO, INPUT OBSERVED SAMPLE MEAN, STANDARD DEVIATION, AND
SAMPLE SIZE.
n
DATA> 28.4 44 42

THE POSTERIOR DENSITY FOR M IS NORMAL
WITH MEAN AND STANDARD DEVIATION:
   28.4000   6.8663

MTB > exec 'm_cont'

DO YOU WISH TO USE A FLAT PRIOR DENSITY FOR M? (TYPE 'y' OR 'n'.)
y

OBSERVED DATA IN WORKSHEET? (TYPE 'y' OR 'n'.)
   IF NO, INPUT OBSERVED SAMPLE MEAN, STANDARD DEVIATION, AND
SAMPLE SIZE.
n
```

```
THE POSTERIOR DENSITY FOR M IS NORMAL
WITH MEAN AND STANDARD DEVIATION:
  3.6000   7.9016
```

The improvement in percentage increase in stair-climbing power of exercise is $d = m_Y - m_N$. The mean of d is 28.4 - 3.6 = 24.8. The standard deviation of d is $\sqrt{6.87^2 + 7.90^2} = 10.47$. Or, using Minitab:

```
MTB > exec 'mm_cont'

INPUT MEAN AND STANDARD DEVIATION FOR NORMAL DISTRIBUTION FOR MEAN
M1:
DATA> 28.4000    6.8663

INPUT MEAN AND STANDARD DEVIATION FOR NORMAL DISTRIBUTION FOR MEAN
M2:
DATA> 3.6000    7.9016

THE POSTERIOR DENSITY FOR M1-M2 IS NORMAL
WITH MEAN AND STANDARD DEVIATION:
  24.8    10.4681
```

(a) To find the probability that d > 0 calculate the z-score: $z = (0 - 24.8)/10.47 = -2.37$. According to the Standard Normal Table the probability to the right of $z = -2.37$, which equals the probability of z to the left of 2.37, is about 99.1%.

(b) The 90% posterior probability interval for d is $24.8 \pm 1.65 \times 10.47 = 24.8 \pm 17.3$, or from about 7.5 to 42.1.

(c) The 95% posterior probability interval for d is $24.8 \pm 1.96 \times 10.47 = 24.8 \pm 20.5$, or from about 4.3 to 45.3. This interval does not contain 0: The null hypothesis is not supported.

12.27 The following table shows the calculations for Temp. O_2 (T) and Control (C):

	m_0	h_0	$\bar{x}$	s	n	k	h	c_0	c	c_1	m_1	h_1
T	30	20	25.9	8	200	1.0005	8.004	.0025	3.1219	3.1244	25.903	.5657
C	30	20	17.6	6	200	1.0005	6.003	.0025	5.55	5.5525	17.606	.4244

Alternatively, using Minitab:

```
MTB > exec 'm_cont'

DO YOU WISH TO USE A FLAT PRIOR DENSITY FOR M? (TYPE 'y' OR 'n'.)
   IF NO, INPUT MEAN AND STANDARD DEVIATION FOR THE PRIOR DENSITY..
n
DATA> 30 20
```

```
OBSERVED DATA IN WORKSHEET? (TYPE 'y' OR 'n'.)
  IF NO, INPUT OBSERVED SAMPLE MEAN, STANDARD DEVIATION, AND
SAMPLE SIZE.
n
DATA> 25.9 8 200

THE POSTERIOR DENSITY FOR M IS NORMAL
WITH MEAN AND STANDARD DEVIATION:
  25.9033    0.5657

MTB > exec 'm_cont'

DO YOU WISH TO USE A FLAT PRIOR DENSITY FOR M? (TYPE 'y' OR 'n'.)
  IF NO, INPUT MEAN AND STANDARD DEVIATION FOR THE PRIOR DENSITY..
n
DATA> 30 20

OBSERVED DATA IN WORKSHEET? (TYPE 'y' OR 'n'.)
  IF NO, INPUT OBSERVED SAMPLE MEAN, STANDARD DEVIATION, AND
SAMPLE SIZE.
n
DATA> 17.6 6 200

THE POSTERIOR DENSITY FOR M IS NORMAL
WITH MEAN AND STANDARD DEVIATION:
  17.6056    0.4244
```

For the difference d:

```
MTB > exec 'mm_cont'

INPUT MEAN AND STANDARD DEVIATION FOR NORMAL DISTRIBUTION FOR MEAN
M1:
DATA> 25.9033    0.5657

INPUT MEAN AND STANDARD DEVIATION FOR NORMAL DISTRIBUTION FOR MEAN
M2:
DATA> 17.6056    0.4244

THE POSTERIOR DENSITY FOR M1-M2 IS NORMAL
WITH MEAN AND STANDARD DEVIATION:
  8.2977    0.7072
```

(a) To find PdAL0, calculate z at d = 0: z = (0 - 8.30)/.707 = -11.73. So PdAL0 is essentially 1.

(b) The answer to this is clear from part (a), but proceeding formally: The 95% probability interval for d is 8.30 ± 1.96x.707 = 8.30 ± 1.39, or from about 6.91 to 9.68. Since d = 0 is not in this interval, the evidence does not support the null hypothesis. Indeed, as the figure below shows, the 95% probability interval is very far from the null hypothesis d = 0.

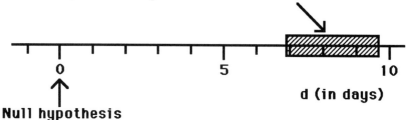

12.29 Using Minitab:

```
MTB > exec 'm_cont'

DO YOU WISH TO USE A FLAT PRIOR DENSITY FOR M? (TYPE 'y' OR 'n'.)
y

OBSERVED DATA IN WORKSHEET? (TYPE 'y' OR 'n'.)
   IF YES, INPUT NUMBER OF COLUMN.
y
DATA> 1

OBS_DATA
   657  623  652  654  658  646  600  640  605  635  642

MEAN
   637.455

STD
   19.2135

COUNT
   11

THE POSTERIOR DENSITY FOR M IS NORMAL
WITH MEAN AND STANDARD DEVIATION:
   637.455      6.751

MTB > exec 'm_cont'

DO YOU WISH TO USE A FLAT PRIOR DENSITY FOR M? (TYPE 'y' OR 'n'.)
y
```

```
OBSERVED DATA IN WORKSHEET? (TYPE 'y' OR 'n'.)
   IF YES, INPUT NUMBER OF COLUMN.
Y
DATA> 1

OBS_DATA
   668   667   647   693   635   644   665   689   642   673   675   641

MEAN
  661.583

STD
  18.6345

COUNT
   12

THE POSTERIOR DENSITY FOR M IS NORMAL
WITH MEAN AND STANDARD DEVIATION:
  661.583      6.126

MTB > exec 'm_cont'

DO YOU WISH TO USE A FLAT PRIOR DENSITY FOR M? (TYPE 'y' OR 'n'.)
Y

OBSERVED DATA IN WORKSHEET? (TYPE 'y' OR 'n'.)
   IF YES, INPUT NUMBER OF COLUMN.
Y
DATA> 1

OBS_DATA
   641   589   603   642   612   603   593   672   612   678   593   602

MEAN
   620

STD
  29.4986

COUNT
   12

THE POSTERIOR DENSITY FOR M IS NORMAL
WITH MEAN AND STANDARD DEVIATION:
  620.000      9.698
```

(**a**) Using the above means and standard deviations in Minitab to find the density of $d = m_3 - m_1$:

```
MTB > exec 'mm_cont'

INPUT MEAN AND STANDARD DEVIATION FOR NORMAL DISTRIBUTION FOR
MEAN M1:
DATA> 661.583     6.126

INPUT MEAN AND STANDARD DEVIATION FOR NORMAL DISTRIBUTION FOR
MEAN M2:
DATA> 637.455     6.751

THE POSTERIOR DENSITY FOR M1-M2 IS NORMAL
WITH MEAN AND STANDARD DEVIATION:
   24.128   9.11613
```

The z-score for $d = m_3 - m_1 = 0$ is $z = (0 - 24.1)/9.12 = -2.65$. According to the Standard Normal Table, the probability to the right of $z = -2.65$—which is the probability to the left of $z = 2.65$—is about 99.6%.

(**b**) Using the above means and standard deviations in Minitab to find the density of $d = m_3 - m_5$:

```
MTB > exec 'mm_cont'

INPUT MEAN AND STANDARD DEVIATION FOR NORMAL DISTRIBUTION FOR
MEAN M1:
DATA> 661.583     6.126

INPUT MEAN AND STANDARD DEVIATION FOR NORMAL DISTRIBUTION FOR
MEAN M2:
DATA> 620.000     9.698

THE POSTERIOR DENSITY FOR M1-M2 IS NORMAL
WITH MEAN AND STANDARD DEVIATION:
   41.583   11.4708
```

The z-score for $d = m_3 - m_5 = 0$ is $z = (0 - 41.6)/11.47 = -3.63$. According to the Standard Normal Table, the probability to the right of $z = -3.63$—which is the probability to the left of $z = 3.63$—is about 99.98%.

12.31 Using Minitab:

```
MTB > exec 'm_cont'

DO YOU WISH TO USE A FLAT PRIOR DENSITY FOR M? (TYPE 'y' OR 'n'.)
y

OBSERVED DATA IN WORKSHEET? (TYPE 'y' OR 'n'.)
  IF NO, INPUT OBSERVED SAMPLE MEAN, STANDARD DEVIATION, AND
SAMPLE SIZE.
n
DATA> 8.27 8.84 11
```

```
THE POSTERIOR DENSITY FOR M IS NORMAL
WITH MEAN AND STANDARD DEVIATION:
  8.27000   3.10592

MTB > exec 'm_cont'

DO YOU WISH TO USE A FLAT PRIOR DENSITY FOR M? (TYPE 'y' OR 'n'.)
y

OBSERVED DATA IN WORKSHEET? (TYPE 'y' OR 'n'.)
  IF NO, INPUT OBSERVED SAMPLE MEAN, STANDARD DEVIATION, AND
SAMPLE SIZE.
n
DATA> 6.17 5.37 12

THE POSTERIOR DENSITY FOR M IS NORMAL
WITH MEAN AND STANDARD DEVIATION:

MEAN STD
  6.17000   1.76549

MTB > exec 'm_cont'

DO YOU WISH TO USE A FLAT PRIOR DENSITY FOR M? (TYPE 'y' OR 'n'.)
y

OBSERVED DATA IN WORKSHEET? (TYPE 'y' OR 'n'.)
  IF NO, INPUT OBSERVED SAMPLE MEAN, STANDARD DEVIATION, AND
SAMPLE SIZE.
n
DATA> 20.67 9.89 12

THE POSTERIOR DENSITY FOR M IS NORMAL
WITH MEAN AND STANDARD DEVIATION:
  20.6700   3.2515
```

(a) Using the above means and standard deviations in Minitab to find the density of $d = m_{Ther} - m_{Comp}$:

```
MTB > exec 'mm_cont'

INPUT MEAN AND STANDARD DEVIATION FOR NORMAL DISTRIBUTION FOR
MEAN M1:
DATA> 8.27000   3.10592

INPUT MEAN AND STANDARD DEVIATION FOR NORMAL DISTRIBUTION FOR
MEAN M2:
DATA> 6.17000   1.76549

THE POSTERIOR DENSITY FOR M1-M2 IS NORMAL
WITH MEAN AND STANDARD DEVIATION:
    2.1   3.57263
```

The 95% probability interval for $d = m_{Ther} - m_{Comp}$ is $2.1 \pm 1.96 \times 3.57 = 2.1 \pm 7.0$, or from about -4.9 to 9.1. Since $d = 0$ is contained in this interval the null hypothesis is supported. (The null hypothesis corresponds to no difference (on the average) between the effectivenesses of the therapist and the computer.)

(**b**) Using the above means and standard deviations in Minitab to find the density of $d = m_{Ther} - m_{Cont}$:

```
MTB > exec 'mm_cont'

INPUT MEAN AND STANDARD DEVIATION FOR NORMAL DISTRIBUTION FOR
MEAN M1:
DATA> 8.27000    3.10592

INPUT MEAN AND STANDARD DEVIATION FOR NORMAL DISTRIBUTION FOR
MEAN M2:
DATA> 20.6700    3.2515

THE POSTERIOR DENSITY FOR M1-M2 IS NORMAL
WITH MEAN AND STANDARD DEVIATION:
   -12.4    4.49655
```

The 95% probability interval for $d = m_{Ther} - m_{Cont}$ is $-12.4 \pm 1.96 \times 4.50 =$ -12.4 ± 8.8, or from about -21.2 to -3.6. Since $d = 0$ is not contained in this interval the null hypothesis is not supported. (The evidence suggests that the therapist is more effective than the control.)

(**c**) Using the above means and standard deviations in Minitab to find the density of $d = m_{Comp} - m_{Cont}$:

```
MTB > exec 'mm_cont'

INPUT MEAN AND STANDARD DEVIATION FOR NORMAL DISTRIBUTION FOR
MEAN M1:
DATA> 6.17000    1.76549

INPUT MEAN AND STANDARD DEVIATION FOR NORMAL DISTRIBUTION FOR
MEAN M2:
DATA> 20.6700    3.2515

THE POSTERIOR DENSITY FOR M1-M2 IS NORMAL
WITH MEAN AND STANDARD DEVIATION:
   -14.5    3.69989
```

The 95% probability interval for $d = m_{Comp} - m_{Cont}$ is $-14.5 \pm 1.96 \times 3.70 =$ -12.4 ± 8.8, or from about -21.8 to -7.2. Since $d = 0$ is not contained in this interval the null hypothesis is not supported. (The evidence suggests that the computer is more effective than the control.)

CHAPTER 13
DATA TRANSFORMATIONS AND
NONPARAMETRIC METHODS

[Just as in Chapters 11 and 12, these answers use the small-n correction factor regardless of the size of n. If you choose not to use this correction when n is large then your answer will be slightly different from mine. Minitab can be used with benefit in this chapter as well as in the previous chapters, but I did not use it. For the one-sample exercises in which calculations as in Chapter 11 are appropriate (when suitably modified), use m_cont. For the two-sample exercises similar to those in Chapter 12, use m_cont and then mm_cont.]

13.1 The following table gives the logs of the number of months at which the children spoke their first word:

Mo.	log	Mo.	log	Mo.	log
15	2.7081	11	2.3979	11	2.3979
26	3.2581	8	2.0794	10	2.3026
10	2.3026	20	2.9957	12	2.4849
9	2.1972	7	1.9459	42	3.7377
15	2.7081	9	2.1972	17	2.8332
20	2.9957	10	2.3026	11	2.3979
18	2.8904	11	2.3979	10	2.3026

We have these values: $m_0 = 2.5$, $h_0 = 1.8$, $\bar{x} = 2.5635$, $s = .4214$, $n = 21$. Using the formulas given in Section 11.3: $k = 1.0454$, $h = .4405$, $c_0 = .3086$, $c = 108.22$, $c_1 = 108.528$, $m_1 = 2.5633$, $h_1 = .096$.

(a) The posterior density of m is normal(2.56, .096).

(b) The predictive distribution is normal $(m_1, \sqrt{h^2 + 1/c_1})$ = normal(2.56, .451). For $x = 3.18$, the z-score is $z = (3.18 - 2.56)/.451 = 1.37$. From the Standard Normal Table, the probability to the right of $z = 1.37$ (which is the same as the probability to the left of $z = -1.37$) is about 8.5%.

13.3 The averages of the two drug days C and D, the four nondrug days A, B, E and F, and also their differences are shown in the accompanying table. Consider only the last column. Its mean is .2314 and its standard deviation is $s = .1843$. Using the formulas in Section 11.3: $k = 1.0889$, $h = .2006$, $c_1 = 372.6$, $m_1 = .2314$, $h_1 = .0518$. So the density of m is normal(.2314, .0518).

(a) The z-score for $m = 0$ is $z = (0 - .231)/.0518 = -4.47$. The probability to the right of this value of z is essentially 1.

(**b**) The 99% posterior probability interval for m is $.231 \pm 2.58 \times .0518 = .231 \pm .134$, or from about .097 to .365. Transforming to percentage increases using antilogs: $e^{.097} = 1.10$ or a 10% increase and $e^{.365} = 1.44$ or a 44% increase. So the experimental drug increases digoxin level by between 10 and 44 percent with high probability.

(**c**) The answer is clear since part (a) indicates that nearly all the probability is to the right of 0. Formally, the 95% posterior probability interval for m is $.231 \pm 1.96 \times .0518 = .231 \pm .102$, or from about .129 to .333. This interval does not contain 0 and so the null hypothesis of no effect (on digoxin level) of the experimental drug is not supported. (Alternatively, transforming to percentage increases using antilogs: $e^{.129} = 1.14$ or a 14% increase and $e^{.333} = 1.40$ or a 40% increase; 0% is not contained in the interval from 14% to 40%.)

	Logs of Plasma digoxin		
Pt#	ave C,D	ave A,B,E,F	Diff.
1	3.916	3.884	.032
2	3.838	3.565	.273
3	4.006	3.544	.462
4	3.428	3.403	.025
5	3.987	4.045	-.057
6	3.919	3.830	.090
7	4.072	3.623	.450
8	4.188	3.972	.216
9	4.121	3.578	.543
10	3.857	3.482	.375
11	4.085	3.829	.257
12	3.636	3.380	.256
13	3.744	3.761	-.017
14	4.158	3.867	.291
15	4.158	3.881	.277

13.5 The following table shows the calculations for the logarithms:

Subj	V	V+	log V	log V+	Diff.
1	2.55	3.15	.936	1.147	.211
2	1.81	2.07	.593	.728	.134
3	1.99	3.22	.688	1.169	.481
4	2.37	2.67	.863	.982	.119
5	3.03	2.90	1.109	1.065	-.044
6	2.25	2.47	.811	.904	.093
7	1.89	1.31	.637	.270	-.367
8	1.83	2.68	.604	.986	.382

Only the last column is relevant for the remainder of this problem. The mean is .1263 and the standard deviation is .2427. Using the formulas in Section 11.3: $k = 1.313$, $h = .3186$, $c_1 = 78.81$, $m_1 = .1263$, $h_1 = .1126$. The posterior density of m (the population mean of the logs of differences) is normal(.1263, .1126). We want to find the probability that m is greater than $\log(1.30) = .2624$. The z-score is $z = (.2624 - .1263)/.1126 = 1.21$. According to the Standard Normal Table the probability to the right of 1.21 is about 11.3%.

13.7 These are the logs:

X yield (g)	X log	Y yield (g)	Y log
1595	7.3746	1520	7.3265
1630	7.3963	1455	7.2828
1515	7.3232	1450	7.2793
1635	7.3994	1480	7.2998
1625	7.3933	1445	7.2759

The original data and the logged data are shown in the accompanying dot plots—the left-hand plot is repeated from the answer to Exercise 12.14. The benefit of taking logs is not clear: the variability in both samples is about the same both with and without taking logs. And transforms are usually designed to pull outliers back into the fold, but the relative distance from the smallest X, say, to the other X's is about the same with and without logs. However, you will see that the answer to this exercise is nearly the same as the answer to Exercise 12.14, and so while there's no obvious benefit in taking logs, nothing is lost.

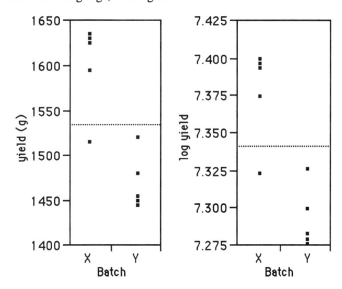

These are the relevant calculations:

	$\bar{x}$	s	n	k	h	c_1	m_1	h_1
X:	7.377	.0284	5	1.8	.0512	1909	7.377	.0229
Y:	7.293	.0187	5	1.8	.0337	4405	7.293	.0151

The density of the difference, $d = m_X - m_Y$, is normal(7.377-7.293, $\sqrt{.0229^2 + .0151^2}$) = normal(.084,.0274).

(a) To find the posterior probability that d > 0, that is, that the mean log of yield for batch X is bigger than that for batch Y, calculate the z-score: $z = (0-.084)/.0274 = -3.08$. (This is nearly identical with $z = -3.07$ found in Exercise 12.14, as advertised above.) The probability to the right of $z = -3.08$ is about 99.9%.

(b) The 95% probability interval for d is $.084 \pm 1.96 \times .0274 = .084 \pm .054$, or from about .030 to .138. Since d = 0 is not contained in this interval, the evidence does not support the null hypothesis. (So the difference between the two samples that is shown in the dot plot is probably real.)

13.9 These are the relevant calculations:

	$\bar{x}$	s	n	k	h	c_1	m_1	h_1
A/C:	2.117	2.658	12	1.1389	3.027	1.310	2.117	.8737
B/C:	.614	2.132	12	1.1389	2.429	2.035	.614	.7011

(a) The posterior density of $m_{A/C}$ is normal(2.117,.8737). "Drug A allows more premature beats than does Drug C" means $m_{A/C} > 0$. The z-score of $m_{A/C} = 0$ is $z = (0-2.117)/.8737 = -2.42$. The probability to the right of this (from the Standard Normal Table) is about 99.2%.

(b) "Drug C cuts by 50%" means difference $m_{A/C}$ is at least $\log(2) = .693$. The z-score for $m_{A/C} = .693$ is $z = (.693-2.117)/.8737 = -1.63$. The probability to the right of $z = -1.63$ is about 94.8%.

(c) The posterior density of $m_{B/C}$ is normal(.614,.7011). The z-score of $m_{B/C} = 0$ is $z = (0-.612)/.7011 = -.87$. The probability to the right of this (from the Standard Normal Table) is about 80.8%.

(d) The z-score for $m_{B/C} = .693$ is $z = (.693-.614)/.7011 = .11$. The probability to the right of $z = .11$ is about 45.6%.

13.11 These are the group numbers for each pot:

Pot	1	2	3	4	5	6	7	8	9	10	11	12	13	14	15
Diff.	49	-67	8	16	6	23	28	41	14	27	56	24	75	60	-48
Group	2	-2	1	1	1	1	1	2	1	1	2	1	2	2	-2

These are the relevant calculations:

$\bar{x}$	s	n	k	h	c_1	m_1	h_1
.933	1.236	15	1.0889	1.346	8.274	.933	.348

(a) The posterior density of m is normal(.933, .348).

(b) The z-score for m = 0 is

$$z = \frac{0 - .933}{.348} = -2.68$$

From the Standard Normal Table the probability of z > -2.68, which is the same as the probability of z < +2.68 is about .996. This is the probability that 'cross' produces more vigorous plants than 'self.'

13.13 These are the differences, Drug A minus Drug C, and the categories of the differences:

Diff, A-C	Cat. of diff.
170	3
13	1
169	3
9	1
194	3
19	1
4	0
469	3
0.4	0
-34.6	-2
1	0
29	1

These are the relevant calculations for the category of differences:

$\bar{x}$	s	n	k	h	c_1	m_1	h_1
1.167	1.518	12	1.1389	1.729	4.013	1.167	.499

To find the posterior probability that Drug C is more effective than Drug A, calculate the z-score for m = 0: z = (0-1.167)/.499 = -2.34. The probability to the right of this value of z is about 99.0%.

13.15 These are the relevant calculations for the differences of categories, A-C:

$\bar{x}$	s	n	k	h	c_1	m_1	h_1
1.167	1.675	12	1.1389	1.908	3.298	1.167	.551

To find the posterior probability that Drug B is more effective than Drug A, calculate the z-score for m = 0: z = (0-1.167)/.551 = -2.12. The probability to the right of this value of z is about 98.3%.

13.17 These are the categories, given in the same order as in the exercise:

Oxygen gel: 3, 4, 3, 3, 4   2, 3, 3, 2, 4   3, 3, 3, 3, 3
             2, 3, 4, 4, 2   4, 3, 3, 3, 2   4, 3, 2, 3, 2

Placebo gel: 2, 3, 2, 2, 2   2, 3, 2, 2, 2   1, 3, 1, 2, 2
             2, 2, 3, 1, 2   1, 2, 3, 3, 3   2, 4, 4, 2, 2
             2, 2, 4, 2

These are the relevant calculations for the two categories in groups T and C:

	$\bar{x}$	s	n	k	h	c_1	m_1	h_1
T:	3.000	.683	30	1.022	.698	61.52	3.000	.127
C:	2.265	.779	34	1.017	.792	54.18	2.265	.139

(a) To find PdAL0, the probability that oxygen gel has a greater average effectiveness than placebo gel, calculate the z-score for $d = 0$:

$$z = \frac{0 - (3.000 - 2.265)}{\sqrt{.127^2 + .136^2}} = \frac{-.735}{.186} = -3.94$$

The probability to the right of this value of z is about 99.99%.

(b) The 99% probability interval is $.735 \pm 2.58 \times .186 = .735 \pm .480$, or from about .255 to 1.22.

(c) The interval in part (b) converts to index scores of about 3.7 ± 2.4, or from 1.3 to 6.1.

(d) The 95% probability interval is $.735 \pm 1.96 \times .186 = .735 \pm .365$, or from about .37 to 1.10. Since this does not contain 0, the null hypothesis is not supported (and the conclusion that the oxygen is effective in improving the average of this index is supported).

13.19 These are the categories, given in the same order as in the exercise:

Inhibitor Group					Control Group			
--	0	1	6		0	0	0	0
--	1	1	6		0	0	0	0
4	1	1	6		0	0	0	0
0	1	2	6		4	0	0	0
0	1	6	6		0	0	0	2

These are the relevant calculations for the two categories in groups T and C:

	m_0	h_0	$\bar{x}$	s	n	k	h	c_0	c	c_1	m_1	h_1
T:	2.5	1.42	2.72	2.47	18	1.06	2.62	.50	2.6	3.12	2.69	.566
C:	.3	.70	.30	.95	20	1.05	1.00	2.04	19.9	22.0	.38	.213

To find PdAL0, the probability that the inhibitor rats live longer, calculate the z-score for $d = 0$:

$$z = \frac{0 - (2.69 - .38)}{\sqrt{.566^2 + .213^2}} = \frac{-2.30}{.605} = -3.81$$

The probability to the right of this value of z is about 99.99%.

13.21 These are the ranks, given in the same order as in the exercise:

Exp1	Exp2	Diff	Exp1	Exp2	Diff	Exp1	Exp2	Diff
38	39	-1	20	21	-1	19	22	-3
30	33	-3	12	14	-2	25	23	+2
28	31	-3	16	18	-2	10	11	-1
37	41	-4	8	13	-5	4	9	-5
32	36	-4	15	17	-2	1	2	-1
27	24	+3	40	42	-2	7	6	+1
3	5	-2	26	29	-3	35	34	+1

These are the relevant calculations for the differences in ranks, Exp. 1 minus Exp. 2:

$\bar{x}$	s	n	k	h	c_1	m_1	h_1
-1.76	2.091	21	1.045	2.186	4.396	-1.76	.477

To find the posterior probability that the population mean is negative, find PdAL0 and subtract it from 1. Equivalently, find the probability to the left of the corresponding z-score:

$$z = \frac{0 - (-1.76)}{.477} = \frac{1.76}{.477} = 3.69$$

From the Standard Normal Table, the probability to the left of z = 3.69 is about 99.99%.

13.23 These are the averages of the ranks as indicated in the exercise:

Pt#	Ave.C&D	Ave.A,B,E,F	Diff
1	56.75	41.25	15.5
2	48.5	34.875	13.625
3	66.75	28	38.75
4	17.75	16.875	0.875
5	64.25	52.375	11.875
6	57.75	49.375	8.375
7	71.5	30.25	41.25
8	80	61.375	18.625
9	71	38.75	32.25
10	51.75	22	29.75
11	71.75	48	23.75
12	29.25	16	13.25
13	39	42.25	-3.25
14	78.25	47.75	30.5
15	76.25	53.875	22.375

These are the relevant calculations for the difference in average ranks:

$\bar{x}$	s	n	k	h	c_1	m_1	h_1
19.83	12.63	15	1.089	13.76	.0793	19.83	3.552

85

(a) To find the posterior probability that m > 0, calculate the z-score:

$$z = \frac{0 - 19.83}{3.552} = -5.58$$

(This compares with z = -4.47 you got taking logs in Exercise 13.3. So the conclusion is even stronger that the mean of the difference in ranks is positive than it was that the means of the differences in logarithms is positive. The reason is that ranking pulls in outliers even more than does taking logarithms. Comparing the two accompanying histograms of differences you will see that the one for ranks is a little tighter. The comparison is far from startling because both transformations do pretty well.) The posterior probability to the right of z = -5.58 is essentially 1.

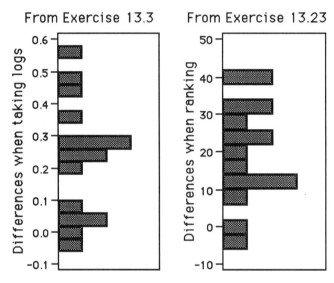

(b) To test the hypothesis that m = 0, find the 95% posterior probability interval: 19.83 ± 1.96x3.552, or 19.83 ± 6.96, which is from 12.87 to 26.79. This interval does not come close to containing 0 and so the null hypothesis that m is 0 is not close to being supported.

13.25 The ranks and desired differences are shown in the accompanying table. These are the relevant calculations for the two columns of differences:

	$\bar{x}$	s	n	k	h	c_1	m_1	h_1
A - C:	9.458	11.61	12	1.139	13.23	.0686	9.458	3.818
B - C:	3.667	8.68	12	1.139	9.88	.1229	3.667	2.852

Ranks and their differences:

Pt#	A	B	C	A-C	B-C
1	32	15.5	2.5	29.5	13
2	20	10.5	14	6	-3.5
3	33	34	19	14	15
4	18	6	9	9	-3
5	35	5	21	14	-16
6	29	26	25	4	1
7	15.5	28	12	3.5	16
8	36	17	13	23	4
9	7	8	2.5	4.5	5.5
10	10.5	30	27	-16.5	3
11	23	31	22	1	9
12	24	2.5	2.5	21.5	0

(a) To find the posterior probability that $m_A > 0$, calculate the z-score:

$$z = \frac{0 - 9.458}{3.818} = -2.48$$

The posterior probability to the right of $z = -2.48$ is about 99.3%.

(b) To test the hypothesis that $m_A = 0$, find the 95% posterior probability interval: $9.458 \pm 1.96 \times 3.818$, or 9.46 ± 7.48, which is from about 1.97 to 16.94. This interval does not contain 0 and so the null hypothesis that m_A is 0 is not supported.

(c) To find the posterior probability that $m_B > 0$:

$$z = \frac{0 - 3.667}{2.852} = -1.29$$

The posterior probability to the right of $z = -1.29$ is about 90.1%.

(d) The 95% posterior probability interval is $3.667 \pm 1.96 \times 2.852$, or 3.67 ± 5.59, which is from about -1.92 to 9.26. This interval contains 0 and so the null hypothesis that m_B is 0 is supported.

13.27 Letting "+" be "success," then in the notation of Chapters 7 and 9: $a = 1$, $b = 1$, $s = 4$, $f = 12$. The posterior distribution of the population proportion in which the resistance is greater on Electrode 1 is $beta(a+s, b+f) = beta(5,13)$.

(a) The predictive probability that Electrode 1 resistance will be greater than that of Electrode 2 on the next subject is $\frac{a+s}{a+s+b+f} = 5/18 = .278$.

(**b**) To carry out a sign test of the null hypothesis that Electrode 1 resistance is greater than that of Electrode 2 for half of the population, find the z-score for population proportion = 1/2. Calculate as follows:

$$r = \frac{a+s}{a+s+b+f} = \frac{5}{18} = .278, \quad r^+ = \frac{a+s+1}{a+s+b+f+1} = \frac{6}{19},$$

$$t = \sqrt{r(r^+ - r)} = \sqrt{.01056} = .103, \quad z = \frac{.5 - r}{t} = \frac{.222}{.103} = 2.16$$

This value of z is not between -1.96 and 1.96. Therefore the null hypothesis is not supported. Equivalently, the 95% probability interval is $5/18 \pm 1.96 \times .103$ $= .278 \pm .202$, or from .076 to .480; this interval does not contain .5.

(Because this conclusion is close and because the sample size is small, an exact calculation (using Minitab, say) might conclude that the null hypothesis is supported. I wouldn't worry too much about this because you should not take "supports" or "does not support" to mean "null hypothesis true" or "null hypothesis not true." This is especially so when 1/2 is close to the end of the 95% probability interval. It happens that the exact 95% probability interval for the beta(5,13) density is from .103 to .499.)

(**c**) For the posterior probability that Electrode 1 resistance is greater than that of Electrode 2 for more than one-third of the population, use z-score:

$$z = \frac{1/3 - r}{t} = \frac{.056}{.103} = .54$$

The probability to the right of z = .54 is about 29.5%. (From Minitab, the exact probability to the right of .5 for the beta(5,13) density is 28.1%.)

13.29 Letting "+" be "success," that is, that the drug gives the greater increase than placebo. Then using the notation of Chapters 7 and 9: a = 3, b = 1, s = 20, f = 10. The posterior distribution of the population proportion in which the drug is more effective than placebo is beta(a+s, b+f) = beta(3+20,1+10) = beta(23,11).

(**a**) The predictive probability that the drug will result in a larger increase than placebo for the next patient is $\frac{a+s}{a+s+b+f} = \frac{23}{34} = .676$.

(**b**) To carry out a sign test to judge whether the increase is greater for drug for half of the population, find the 95% probability interval. Calculate

$$r = \frac{a+s}{a+s+b+f} = \frac{23}{34}, \quad r^+ = \frac{a+s+1}{a+s+b+f+1} = \frac{24}{35}, \quad t = \sqrt{r(r^+ - r)} = .0791$$

The value of z at proportion .5 is z = (.5-.676)/.0791 = -2.23. This does not lie in the interval from -1.96 to 1.96 and so the null hypothesis is not supported. (Equivalently, the approximate 95% probability interval for the beta(23,11) density is $.676 \pm 1.96 \times .0791 = .676 \pm .155$, or from about .521 to .831. This interval does not contain .5. (Using Minitab, the exact 95% probability interval is 51.3% to 82.0%.)

13.31 Let "+" be "success," that is, that the first indicated therapy is better than the second, and use the notation of Chapters 7 and 9:

(a) For formoterol over salbutamol, $a = 1$, $b = 1$, $s = 30$, $f = 0$. The posterior distribution of the population proportion with a better response on F than on S for the next patient from this population is beta(a+s, b+f) = beta(31,1). The predictive probability that F will be greater than S for the next patient is

$$\frac{a+s}{a+s+b+f} = 31/32 = 96.9\%.$$

(b) To carry out a sign test to judge whether salbutamol is more effective than placebo for half of the population (that is, has the same effect as placebo in the sense that half respond better on one and the other half responds better on the other), find the 95% probability interval. For $a = 1$, $b = 1$, $s = 27$, $f = 2$, the posterior density is beta(a+s, b+f) = beta(28, 3). Calculate

$$r = \frac{28}{31} = .9032, \quad r^+ = \frac{29}{32}, \quad t = \sqrt{r(r^+ - r)} = .0550$$

The 95% probability interval for the beta(28, 3) density is, approximately, $.903 \pm 1.96 \times .055 = .903 \pm .108$, or from about .795 to 1.011. (That the upper endpoint is greater than 1 indicates that the approximation is not very good—the reason is that the beta(28, 3) is not close to being symmetric. The exact interval from Minitab is .779 to .979.) This interval does not contain .5 and so the null hypothesis is not supported.

(c) The posterior density for the proportion of patients who show a better response on formoterol than salbutamol is beta(a+s, b+f) = beta(1+25, 1+4) = beta(26, 5). To approximate the posterior probability that this proportion is greater than 80%, calculate the z-score:

$$r = \frac{26}{31} = .839, \quad r^+ = \frac{27}{32},$$

$$t = \sqrt{r(r^+ - r)} = .0710, \quad z = \frac{.8 - r}{t} = \frac{-.0387}{.0710} = -.55$$

The probability to the right of $z = -.55$ is about 70.9%. (The exact probability to the right of .8 for the beta(26, 5) density is 74.5%.)

CHAPTER 14
REGRESSION ANALYSIS

14.1 **(a)** Use the following calculations:

#	x	y	x^2	y^2	xy
1	20	89	400	7921	1780
2	16	72	256	5184	1152
3	20	93	400	8649	1860
4	18	84	324	7056	1512
5	17	81	289	6561	1377
6	16	75	256	5625	1200
7	15	70	225	4900	1050
8	17	82	289	6724	1394
9	15	69	225	4761	1035
10	16	83	256	6889	1328
ave:	17	79.8	292	6427	1368.8

$$\bar{x} = 17, \ s_x = \sqrt{292 - 17^2} = \sqrt{3} = 1.732;$$

$$\bar{y} = 79.8, \ s_y = \sqrt{6427 - 79.8^2} = \sqrt{58.96} = 7.679$$

$$r = \frac{\overline{xy} - \bar{x}\,\bar{y}}{s_x s_y} = \frac{1368.8 - (79.8)(17.0)}{(7.679)(1.732)} = .9173$$

(b) $B = r\dfrac{s_y}{s_x} = .9173\dfrac{1.732}{7.679} = .2069$

$A = \bar{y} - B\,\bar{x} = 17.0 - (.2069)(79.8) = .4878$

Least-squares line is y = .4878 + .2069x

(c) The scatterplot, including the least-squares line, is shown at the top of the next page.

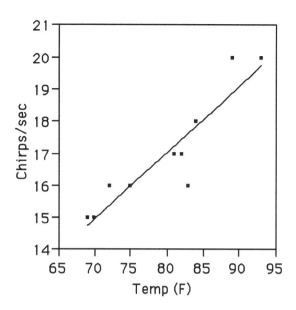

14.3 Using Minitab:

```
MTB > set c1
DATA> .45 0 0 .24 .12 0 .35 .44
DATA> end
MTB > set c2
DATA> 1479 1500 516 1815 1118 1702 1348 1337
DATA> end
MTB > exec 'lin_reg'

INPUT COLUMN NUMBERS OF X AND Y DATA:
DATA> 1 2

THE LEAST-SQUARES LINE HAS SLOPE AND INTERCEPT:
 ROW        B          A
   1     489.246    1254.03

THE POSTERIOR DENSITY FOR b IS NORMAL
WITH MEAN AND STANDARD DEVIATION:
 ROW      MEAN        STD
   1     489.246    920.166

FOR PREDICTING Y FOR GIVEN VALUES OF X, INPUT X VALUES OF INTEREST:
DATA> end
```

So $y = 1254 + 489x$ where y is osmolarity in mg/l and x is rainfall in inches.

Or $y = 1.254 + .489x$ where y is osmolarity measured in g/l.

The scatterplot, including the least-squares line, is shown at the top of the next page.

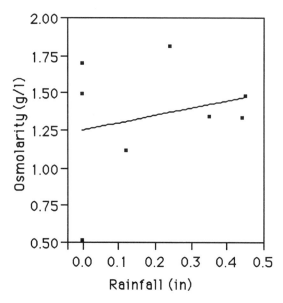

14.5 Least-squares line: (Away Wins) = -1.569 + .6434(Home Wins)

This line is added to figure given in the exercise:

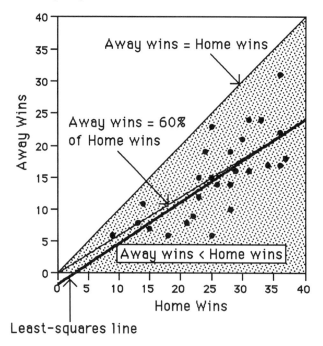

14.7 **(a)** Correlation for 1988: r = .235

Least-squares line: Winning% = .4402 + .000516(Payroll)

(b) Correlation for 1992: r = .0188

Least-squares line: Winning% = .4961 + .000013(Payroll)

Scatterplots (with least-squares lines) on the same scale are shown below:

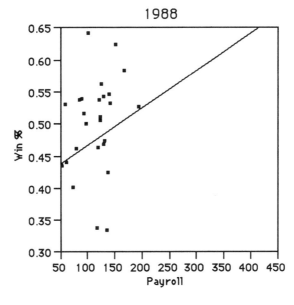

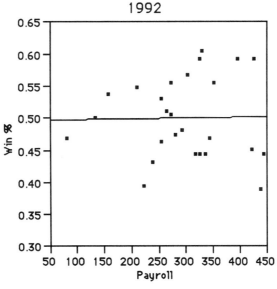

[As an indication of the very slight relationship between these quantities in both years, extrapolate back to Payroll = 0. In 1988 the predicted winning percentage is 44% and in 1992 it is 49.6%, essentially half the games with no payroll at all! Of course, extrapolating so far from the data is risky and in this instance it leads to a clearly silly answer. We don't really believe the relationship continues to be linear for small payrolls. (Nor in 1988 do we believe that a team would have won 65% of its games had its payroll been $40,000,000.) However, such a large value of the least-squares line at 0 does show that there is little relation between winning percentage and payroll.]

14.9 Least squares line: Rad = 8.55 + 1.89(El)

Least-squares line on scatterplot of the data:

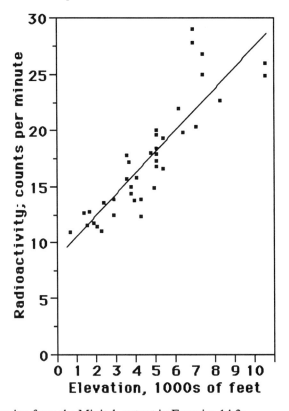

14.11 Recall the following from the Minitab output in Exercise 14.3:

```
THE POSTERIOR DENSITY FOR b IS NORMAL
WITH MEAN AND STANDARD DEVIATION:
  ROW     MEAN        STD
   1    489.246    920.166
```

(a) Assuming flat prior densities for a and b, the 95% probability interval for the slope b is 489 ± 1.96x920 = 489 ± 1804, or from about -1310 to 2290. This is measuring osmolarity in mg/l. Measuring in g/l the 95% probability interval is about -1.31 to 2.29. The figure below is duplicated from Exercise 14.3. Now, in addition to the least-squares line with slope = .489, it contains a line with slope = -1.31 and another with slope = 2.29. These lines are drawn to intersect at the mean rainfall of .2 inches; they are representative of lines with these slopes but there are many other candidate lines.

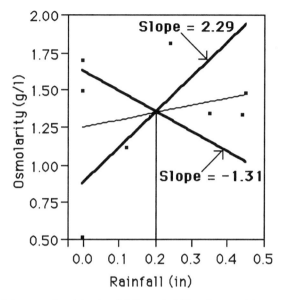

(**b**) Since b = 0 is contained in the 95% probability interval in part (a), the null hypothesis is no relationship is supported.

14.13 Using Minitab:

```
MTB > set c1
DATA> 24 20 16 12 8 6 2 24 20 16 12 8 6 2
DATA> end
MTB > set c2
DATA> -.85 -1.23 -1.38 -1.58 -1.43 -1.71 -1.86 -.81 -1.23 -1.32
-1.53 -1.52 -1.61 -1.79
DATA> end
MTB > exec 'lin_reg'

INPUT COLUMN NUMBERS OF X AND Y DATA:
DATA> 1 2

THE LEAST-SQUARES LINE HAS SLOPE AND INTERCEPT:
  ROW        B         A
   1   0.0390175  -1.90836

THE POSTERIOR DENSITY FOR b IS NORMAL
WITH MEAN AND STANDARD DEVIATION:
  ROW       MEAN        STD
   1     0.0390175  0.0040508
```

(**a**) The least-squares line is y = -1.91 + .039x and is shown in the figure on the next page.

(**b**) The 99% posterior probability interval for the slope of the true line is .039 ± 2.58x.00405 = .039 ± .010, or from about .029 to 0.049.

(**c**) To test the hypothesis that the slope b = 0, check whether b = 0 is in the 95% posterior probability interval. The 95% interval is contained in the 99% interval, as found in part (b). The value b = 0 is not contained in the 99% interval and therefore it cannot be in the 95% interval. [The 95% probability interval is from about .031 to .047.] So the null hypothesis is not supported.

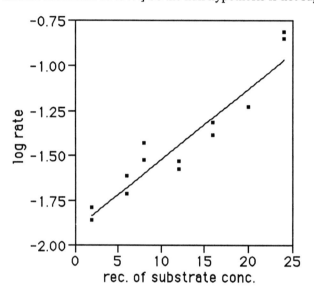

14.15 (a) Finding the distribution of b using Minitab:

```
MTB > set c1
DATA> 67.33 69.33 68.42 68.42 68.75 69.67 70.42 71.8 67.17
DATA> end
MTB > set c2
DATA> 452.99 443.22 412.18 289.66 567.79 621.45 473.64 697.14
288.48
DATA> end
MTB > exec 'lin_reg'
MTB > echo

INPUT COLUMN NUMBERS OF X AND Y DATA:
DATA> 1 2

THE LEAST-SQUARES LINE HAS SLOPE AND INTERCEPT:
  ROW       B        A
   1   69.884   -4352.57

THE POSTERIOR DENSITY FOR b IS NORMAL
WITH MEAN AND STANDARD DEVIATION:
  ROW     MEAN       STD
   1   69.884    26.3616
```

The least-squares line is shown in the accompanying figure. A 95% probability interval for slope b is 69.9 ± 1.96x26.4 = 69.9 ± 51.7, or from about 18.2 to 121. So costs are going up at rate that is probably between 18.2 and 121 million dollars per year.

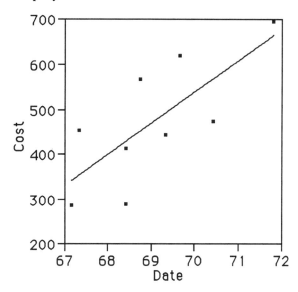

(b) Finding the distribution of b using Minitab:

```
MTB > set c1
DATA> 1065 1065 530 530 913 786 538 1130 821
DATA> end
MTB > set c2
DATA> 452.99 443.22 412.18 289.66 567.79 621.45 473.64 697.14
288.48
DATA> end
MTB > exec 'lin_reg'

INPUT COLUMN NUMBERS OF X AND Y DATA:
DATA> 1 2

THE LEAST-SQUARES LINE HAS SLOPE AND INTERCEPT:
 ROW          B          A
  1    0.277597    244.271

THE POSTERIOR DENSITY FOR b IS NORMAL
WITH MEAN AND STANDARD DEVIATION:
 ROW        MEAN         STD
  1     0.277597    0.206918
```

The least-squares line is shown in the figure on the next page. A 95% probability interval for slope b is .278 ± 1.96x.207 = .278 ± .406, or from about -.13 to .68 (in millions of dollars per unit capacity). Since this interval contains b = 0, the null hypothesis is supported.

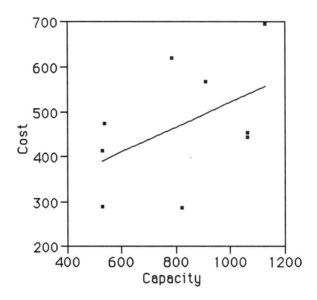

14.17 Using Minitab:

```
MTB > set c1
DATA> .916 .916 .916 1.099 1.253 2.230 2.293 2.398 2.398 2.398 2.773
3.219 3.219
DATA> end
MTB > set c2
DATA> -.942 -1.386 -1.609 -1.022 -1.386 -2.564 -1.966 -1.715 -2.375
-2.526 -2.957 -2.976 -1.897
DATA> end
MTB > exec 'lin_reg'

INPUT COLUMN NUMBERS OF X AND Y DATA:
DATA> 1 2

THE LEAST-SQUARES LINE HAS SLOPE AND INTERCEPT:
 ROW        B          A
  1   -0.632987  -0.680432

THE POSTERIOR DENSITY FOR b IS NORMAL
WITH MEAN AND STANDARD DEVIATION:
 ROW      MEAN       STD
  1   -0.632987  0.145826

FOR PREDICTING Y FOR GIVEN VALUES OF X, INPUT X VALUES OF INTEREST:
DATA> 2.303
DATA> end

THE PREDICTIVE DENSITY OF THE NEXT OBSERVATION FOR DIFFERENT
VALUES OF X IS NORMAL WITH MEAN AND STANDARD DEVIATIONS GIVEN BELOW:
 ROW      X    MEAN_Y      STD_Y
  1    2.303  -2.1382    0.458069
```

(a) The least-squares line is y = -.680 - .633x. (This line is shown in the figure for part (c).)

(b) Doubling the frequency means increasing the log frequency by log(2) = .693, which changes log displacement by an estimated -.633x.693 = -.439. The antilog of -.439 is .645, which means the displacement decreases by an estimated 35.5%. To see that this decrease seems reasonable, consider the following plot of the antilogged least-squares line in the original frequency/displacement scales:

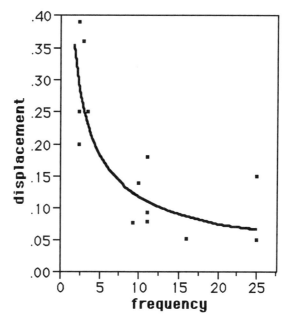

(c) The predictive density of log displacement for a log frequency of 2.303 is normal(-2.14, .458). The z-score of a log displacement of -1.609 is

$$z = \frac{-1.609 - (-2.14)}{.458} = 1.15$$

The probability that this particular house withstands a blast with frequency 10 cps is the probability to the left of z = 1.15, which according to the Standard Normal Table is about 88%. The region (in the log scales) with displacement less that .2 inches and frequency = 10 cps is shown in the figure on the next page.

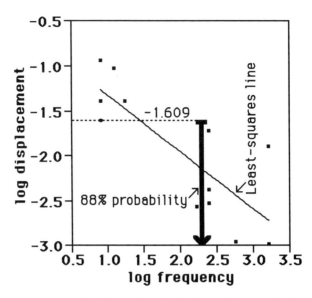

14.19 There are 10 to 20 U.S. major league baseball players in any single year who are capable of hitting many home runs. It is not surprising that some of them do better than Maris's record pace early in the season. This reflects the variability in the process involved in hitting home runs—an occasional player will happen to do well early on. The ones who do well early are good but they are also lucky; they will still be good in the latter part of the season, but they will tend to not be as lucky. That is, they regress toward the mean.

14.21 The next page contains a scatterplot of the Day 3 score and the difference, Day 4 minus Day 3. The correlation coefficient between these two is -.63. The regression effect holds since this is negative. For example, players with high scores on Day 3 tend have negative differences: lower scores on Day 4.

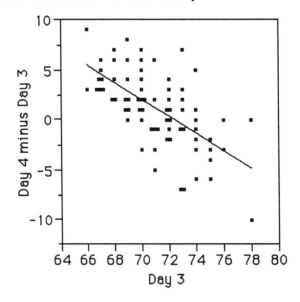

14.23 Using Minitab:

```
MTB > set c1
DATA> 35 34 23 23 30 31 28 31 23 30 31 32 23 14 32 31 26 28 34 25
32 14 13 26 25 29 20 27 38 28
DATA> end
MTB > set c2
DATA> -3 -6 -1 -10 -6 -7 -6 -17 -1 -4 -3 -11 -7 +2 -1 -3 -4    0
-10 -1 +1 -4 +2 -2 -3 -10 +6 -5 -12 -1
DATA> end
MTB > exec 'lin_reg'

INPUT COLUMN NUMBERS OF X AND Y DATA:
DATA> 1 2

THE LEAST-SQUARES LINE HAS SLOPE AND INTERCEPT:
 ROW            B          A
  1    -0.388648    6.33788

THE POSTERIOR DENSITY FOR b IS NORMAL
WITH MEAN AND STANDARD DEVIATION:
 ROW         MEAN        STD
  1    -0.388648    0.128007
```

The least-squares line for y = S - F is y = 6.34 - .389F. If F = 16.3 then this equals 0 (see figure below). If F is less than 16.3 ml then this is positive. But this does not mean that salbutamol should be given to patients who do poorly on formoterol. FEV varies within the same patient from one time to the next. The regression effect means that patients with very low FEV at one time will improve at the next, perhaps even with no treatment. It may be that salbutamol is helping these patients, but it is getting a boost from the regression effect. These patients might have done even better had they been kept on formoterol.

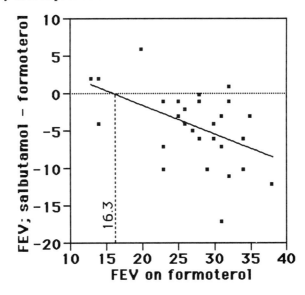

14.25 From Minitab:

```
MTB > set c1
DATA> 25 29 31 40 21 35 23 52 30 45 45 29 39 42 17 35 33 25 50 34 35
55 38 28 31 36 39 50
DATA> end
MTB > set c2
DATA> 27 31 44 43 45 48 44 50 29 34 50 39 50 46 31 36 51 20 57 25 36
55 36 51 41 46 36 39
DATA> end
MTB > exec 'lin_reg'

INPUT COLUMN NUMBERS OF X AND Y DATA:
DATA> 1 2

THE LEAST-SQUARES LINE HAS SLOPE AND INTERCEPT:
ROW        B         A
  1    0.481176   23.6669

THE POSTERIOR DENSITY FOR b IS NORMAL
WITH MEAN AND STANDARD DEVIATION:
ROW      MEAN        STD
  1    0.481176   0.166623

FOR PREDICTING Y FOR GIVEN VALUES OF X, INPUT X VALUES OF INTEREST:
DATA> 20
DATA> end

THE PREDICTIVE DENSITY OF THE NEXT OBSERVATION FOR DIFFERENT
VALUES OF X IS NORMAL WITH MEAN AND STANDARD DEVIATIONS GIVEN BELOW:
ROW    X    MEAN_Y     STD_Y
  1   20    33.2904   8.85466
```

(a) The least-squares line is $y = 23.7 + .481x$, where x = number of sacks in 1991 and y = number of sacks in 1992. (See figure in part (b).)

(b) The predictive density for the number of sacks in 1992 for a team with 20 sacks in 1991 is normal(33.3, 8.85). The z-score for 20 is

$$z = \frac{20 - 33.3}{8.85} = -1.50$$

The probability that this team will have 20 or more sacks in the next year (the region shown by the solid bar in the figure on the next page) is the probability of a z-score greater than -1.50, which according to the Standard Normal Table is about 93%. It is very likely that such a team will improve.

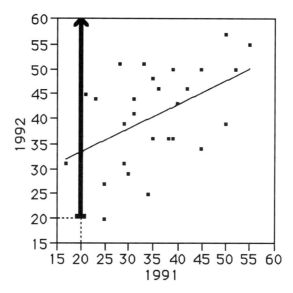

(c) The regression effect would say in this problem that a team with an extreme number of sacks in 1991 (relatively large or relatively small) will tend to have a less extreme number of sacks in 1992. The answer to part (b) demonstrates the regression effect for this problem: a team with 20 sacks, a relatively small number, will likely have more sacks in the following year. Also, the figure below shows the plot of numbers of sacks in 1992 minus 1991 vs. number in 1991; the negative relationship demonstrates the regression effect.

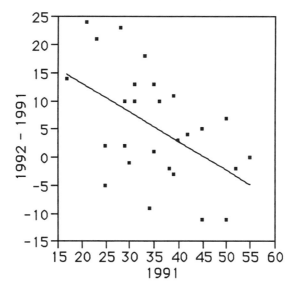

14.27 From Minitab:

```
MTB > set c1
DATA> 25 29 31 40 21 35 23 52 30 45 45 29 39 42 17 35 33 25 50 34 35
55 38 28 31 36 39 50
DATA> end
MTB > set c2
DATA> 34 27 37 46 22 48 34 46 43 46 52 21 35 45 35 29 45 34 51 41 32
36 42 32 44 31 29 31
DATA> end
MTB > exec 'lin_reg'

INPUT COLUMN NUMBERS OF X AND Y DATA:
DATA> 1 2

THE LEAST-SQUARES LINE HAS SLOPE AND INTERCEPT:
ROW         B        A
  1     0.405654   23.0568

THE POSTERIOR DENSITY FOR b IS NORMAL
WITH MEAN AND STANDARD DEVIATION:
ROW       MEAN          STD
  1     0.405654    0.149712

FOR PREDICTING Y FOR GIVEN VALUES OF X, INPUT X VALUES OF INTEREST:
DATA> 20
DATA> end

THE PREDICTIVE DENSITY OF THE NEXT OBSERVATION FOR DIFFERENT
VALUES OF X IS NORMAL WITH MEAN AND STANDARD DEVIATIONS GIVEN BELOW:
ROW      X     MEAN_Y      STD_Y
  1     20    31.1699     7.95598
```

(a) The least-squares line is y = 23.1 + .406x, where x = number of sacks in 1991 and y = number of sacks in 1993. (See figure in part (b).)

(b) The predictive density for the number of sacks in 1992 for a team with 20 sacks in 1991 is normal(31.2, 7.96). The z-score for 20 is

$$z = \frac{20 - 31.2}{7.96} = -1.40$$

The probability that this team will have 20 or more sacks in the next year (the region shown by the solid bar in the figure on the top of the next page) is the probability of a z-score greater than -1.40, which according to the Standard Normal Table is about 92%. It is very likely that such a team will improve over the next two years.

(c) The regression effect would say in this problem that a team with an extreme number of sacks in 1991 (relatively large or relatively small) will tend to have a less extreme number of sacks in 1993. The answer to part (b) demonstrates the regression effect for this problem: a team with 20 sacks, a relatively small number, will likely have more sacks two years hence. Also, the second figure on the next page shows the plot of numbers of sacks in 1993 minus 1991 vs. number in 1991; the negative relationship demonstrates the regression effect.

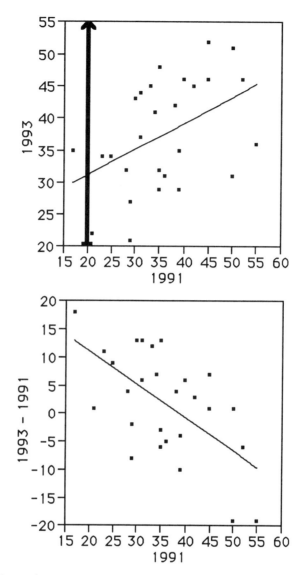

14.29 Depends on the student.